MELANIE

Cross System
Product Application
Development

Books from QED

Database

Managing IMS Databases
Building the Data Warehouse
Migrating to DB2
DB2: The Complete Guide to Implementation and Use
DB2 Design Review Guidelines
DB2: Maximizing Performance of Online Production Systems
Embedded SQL for DB2
SQL for DB2 and SQL/DS Application Developers
How to Use ORACLE SQL*PLUS
ORACLE: Building High Performance Online Systems
ORACLE Design Review Guidelines
Developing Client/Server Applications in an Architected Environment

Systems Engineering

Software Configuration Management
On Time, Within Budget: Software Project Management Practices and Techniques
Information Systems Architecture: Development in the 90's
Quality Assurance for Information Systems
User-Interface Screen Design: Workstations, PC's, Mainframes
Managing Software Projects
The Complete Guide to Software Testing
A Structured Approach to Systems Testing
Rapid Application Prototyping
The Software Factory
Data Architecture: The Information Paradigm
Software Engineering with Formal Metrics
Using CASE Tools for Practical Management

Management

Developing a Blueprint for Data, Applications, and Technology: Enterprise Architecture Planning
Introduction to Data Security and Controls
How to Automate Your Computer Center
Controlling the Future
The UNIX Industry
Mind Your Business

IBM Mainframe Series

From Mainframe to Workstations: Offloading Application Development
VSE/SP and VSE/ESA: A Guide to Performance Tuning
CICS: A Guide to Application Debugging
CICS Application and System Programming
CICS: A Guide To Performance Tuning
MVS COBOL II Power Programmer's Desk Reference
VSE JCL and Subroutines for Application Programmers
VSE COBOL II Power Programmer's Desk Reference
Introduction to Cross System Product
CSP Version 3.3 Application Development
The MVS Primer
MVS/VSAM for the Application Programmer
TSO/E CLISTs: The Complete Tutorial and Desk Reference
CICS: A How-To for COBOL Programmers
QMF: How to Use Query Management Facility with DB2 and SQL/DS
DOS/VSE JCL: Mastering Job Control Language
DOS/VSE: CICS Systems Programming
VSAM: Guide to Optimization and Design
MVS/JCL: Mastering Job Control Language
MVS/TSO: Mastering CLISTs
MVS/TSO: Mastering Native Mode and ISPF
REXX in the TSO Environment, 2nd Edition

Technical

Rdb/VMS: Developing the Data Warehouse
The Wonderful World of the AS/400: Architecture and Applications
C Language for Programmers
Mainframe Development Using Microfocus COBOL/2 Workbench
AS/400: A Practical Guide to Programming and Operations
Bean's Index to OSF/Motif, Xt Intrinsics, and Xlib Documentation for OSF/Motif Application Programmers
VAX/VMS: Mastering DCL Commands and Utilities
The PC Data Handbook
UNIX C Shell Desk Reference
Designing and Implementing Ethernet Networks
The Handbook for Microcomputer Technicians
Open Systems

QED books are available at special quantity discounts for educational uses, premiums, and sales promotions. Special books, book excerpts, and instructive materials can be created to meet specific needs.

This is Only a Partial Listing. For Additional Information or a Free Catalog contact
QED Information Sciences, Inc. • P. O. Box 812070 • Wellesley, MA 02181-0013
Telephone: 800-343-4848 or 617-237-5656 or fax 617-235-0826

Cross System Product Application Development

John King

QED Publishing Group
Boston • Toronto • London

QED Technical Publishing Group is a division of QED Information Sciences, Inc.

Library of Congress Catalog Number: 92-15680
International Standard Book Number: 0-89435-427-2

Printed in the United States of America
93 94 95 10 9 8 7 6 5 4 3 2 1

Library of Congress Cataloging-in-Publication Data

King, John Jay, 1955–
    Cross system product application development / John King.
        p.   cm.
    Includes index.
    ISBN 0-89435-427-2   :
    1. CSP (Computer program language)   I. Title.
QA76.73.C75K56   1992
005.13'3—dc20   92-15680
                                                                        CIP

Contents

Preface

IBM's Cross System Product (CSP) has been around for quite some time, but only recently has interest really started to grow. The reasons for this are two-fold: CSP's status as the official SAA (Systems Application Architecture) Application Generator has aided it greatly, and the combination of significant performance improvements in CSP coupled with rapidly improving hardware to optimize the application developer's time. Reducing time spent both coding and maintaining systems is a major goal in most organizations today. CSP allows huge time savings both when developing well-designed systems and, later, when modifying them.

CSP stands alone as an IBM-supplied application development tool portable across all of IBM's major platforms (TSO, CICS, IMS, AS/400, VM, OS/2, PC-DOS, and others). The ability to generate applications directly from a design specification with a minimum of coding is a hallmark of CSP. Many new systems created for the mainframe environment are using CSP to trim application development costs and time. The time is right for CSP.

This book will show those with a data processing background how to use CSP to create, test, and execute applications. Most technical terms are explained as they are used, but some familiarity with data processing and programming is assumed of the reader. Each chapter is followed by review questions and exercises. The questions will help you know how well the major points from the chapter have "sunk in" and the exercises will provide practice using the material covered in the chapter. By following the chapters in a systematic fashion, you will be led step-by-step through the creation, testing, and use of CSP applications.

Three ways to use this book:

1. If you are new to CSP, scan the chapters quickly, then begin at Chapter 1. Each chapter builds on previous chapters so that you will be able not just to create applications with CSP, but to understand what you are doing.
2. If you have a CSP background already, scan the table of contents and decide which chapters will provide the most valuable information. Then, consider if other chapters might contain some "glue" necessary to hold together other portions of your knowledge.
3. This book should make a good reference for future use.

This book focuses on Version 3 Release 3 of CSP. Users of earlier versions will find this useful, but some features might not be available under the earlier software.

ACKNOWLEDGMENTS

I wish to thank my lovely wife Peggy and my sons Sean and Brian for giving meaning to life (and for putting up with me while this work progressed).

TRADEMARKS

Each term used in this book known (or suspected) to be a trademark of International Business Machines (IBM®) is listed below. I regret any omissions.

AD/CYCLE, AS/400, CICS, CLIST, CPI, CROSS SYSTEM PRODUCT, CSP, CSP/AD, CSP/AE, CUA, DB/2, Dialog Manager, DL/I, DOS, GDDM, IMS, ISPF, MVS, OS/2, OS/400, OS/VS COBOL, PDF, PL/I, REXX, SAA, SDF II, System/370, TSO, VM, VS COBOL II, VSAM, and VSE.

Introduction to Application Development

IBM's Cross System Product (CSP) is the Application Generator portion of IBM's AD/CYCLE (a set of standards governing the Application Development life cycle, encompassing many tools from IBM and non-IBM vendors as well). Solving data processing problems takes much less time when CSP is used than was previously required using conventional programming methods. Applications developers create programs that follow their designs *without* becoming enmeshed in the complexities of second- and third-generation programming tools. However, CSP is *not* a user-friendly 4GL (fourth-generation language). CSP is a product designed to allow the data processing professional to quickly and accurately bring his or her applications to life.

Starting from a good design, CSP application developers can produce systems at many times the rate of programmers using traditional tools. In fact, for a large percentage of applications, CSP may allow developers to produce working applications eight to ten times faster than with traditional programming methods. This efficiency comes from several factors, including good design and reuse of application objects. With CSP the system does more of the work, freeing the programmer for more development. This is not without cost—all of the work must still be done somewhere. It is simply a trade-off, with quicker development being swapped for more complex internal processing and, thus, greater cost for hardware and software. Yet, CSP is not a "silver bullet." There are some things that will still be best solved by creating an application using the old methods. CSP has overhead that may make it unacceptable for applications that are extremely busy or large. In addition, CSP is not an ideal platform for certain types of calculations and data manipulation, and calls to 3GL (third-generation languages such as COBOL, PL/I, and FORTRAN) code may be necessary to get the job done. Overall, CSP is an excellent product for the majority of an average system's online needs.

CSP consists of two major components, CSP Application Development (CSP/AD) and CSP Application Execution (CSP/AE). CSP/AD provides the tools needed to create and test applications. Once the developer is satisfied, the application may be GENERATED to run for one of several CSP/AE environments. This flexibility to develop on one platform and execute on others is a major strength of CSP. The following environments support CSP in one way or another:

MVS/TSO
CICS (both MVS and VSE)
IMS/ESA TM (IMS-DC)
CMS
OS/400
MS-DOS and OS/2
DPPX

As noted before, CSP is a part of IBM's System Application Architecture (SAA). SAA provides a set of standards encompassing most facets of data processing. These standards are intended to culminate in applications that are portable between IBM mainframes (S/370 large-scale systems), AS/400's (midrange systems), and the PS/2 (personal computer) environments. These standards have been designed to vary as the computing world evolves; nothing is "cast in stone."

IBM has crafted standards for Common User Access (CUA), Common Programming Interface (CPI), and Common Communications Support (CCS). CSP is directly impacted by two of these standards, Common User Access and Common Programming Interface.

Common User Access (CUA) defines the interface between the human and the computer. It encompasses multiple definitions attempting to create a similar "look and feel" no matter where you happen to use a computer system. Screen format standards are defined for both Programmable Workstations (PWS) and Non-Programmable Terminals (NPT). These standards cover interface elements, panel appearance, panel-to-panel navigation, color and presentation, and help and message management. The goal is to provide computer system users with a consistent interface so that they know what to expect when they sit down to use a system. This will lower training costs and help make systems more error free. CUA's Programmable Workstation standard brings IBM into the Graphic User Interface (GUI) world. This interface is similar to many now available in the computer world and is growing in sophistication every year. Please note that whether a Programmable Workstation or a Non-Programmable Terminal is being used, the basic format of the panel is easy to recreate. As the power of Programmable Workstations increases, the CUA standard changes to incorporate the latest capabilities including standards for animation and display of high-resolution images. This highlights an ever-widening gap between the Non-Programmable Terminal and the Programmable Workstation. In fact, the Non-Programmable Terminal standard is fairly stabilized due to hardware limitations and will probably not change much as time progresses. Current systems will not be "orphaned." Non-Programmable Terminal users will still have the advantage of a consistent interface, though they will not enjoy the GUI's flexibility (or prettiness).

Common Programming Interface (CPI) defines the interface between system services and programming languages. CPI will provide the basic tools necessary to develop an application. CPI supports several languages and services, most notably for our purposes an Application Generator named Cross System Product (CSP). Besides providing an easier mechanism for the creation of applications, CSP is used as the construction tool for several CASE products. CSP is an integral part of the AD/CYCLE concept.

CSP/AD (APPLICATION DEVELOPMENT)

CSP/AD provides the tools necessary for application developers to create, test, and generate applications. Programmers using CSP/AD can develop software with much less effort than is required using more traditional tools. The shared library capability of CSP allows applications with well-defined standards and complete development plans to save a great deal of time by reusing definitions and code. This reusability can also trim maintenance time greatly. CSP/AD provides facilities for naming libraries and the concatenation order to be followed so that the order of libraries may be altered as needed. CSP/AD provides tools for data definition, map definition, and application definition. CSP/AD is not available on all platforms, most notably IMS. The CSP Programmable Workstation Interface under OS/2 allows definition only; testing and generation must be performed in some other environment.

Most of this book is devoted to teaching you how to use CSP/AD to create applications. Specific chapters are designed to teach you how to define maps, records, and application processes. You will also learn how to test the applications and prepare them for execution. The focus of this book and its illustrations is on TSO, and CICS, both MVS products where CSP is widely used. Throughout the text, notations are made concerning other environments and system dependencies where they are important or differ between environments. One of the strengths of CSP is its sameness in the different environments that support it, so your use of tools will not vary significantly from what is discussed here.

CSP/AE (APPLICATION EXECUTION)

With minor exceptions (discussed in later chapters), CSP applications work the same way under the CSP/AD test facility as they will when later generated and moved to CSP/AE environment. This allows CSP developers to concentrate on creating the application and not waste time worrying about the differences between the test and execution environments. The CSP/AE execution environment provides a stable platform from which to run CSP applications.

CSP applications do not operate as stand-alone executable

modules (except when using CSP/370 Runtime Services in MVS batch and IMS/ESA). A series of CSP/AE routines are necessary at execution time to make them work. In most environments, CSP applications are generated in the CSP/AD environment for execution using CSP/AE in specific execution environments. The generation process creates several members in special libraries called Application Load Files (ALFs). Members of Application Load Files represent code that may be transported to most CSP/AE environments (though the generation must be targeted appropriately). It is important to understand that the ALFs are not executable libraries. The members needed to execute an application are incomplete and must be supplemented by the CSP/AE runtime modules in order for execution to occur. The exceptions to this are the IMS and MVS batch environments where CSP/AD (running somewhere besides IMS) is used to generate VS COBOL II code that is then compiled and linked as a regular load module. Still, the compiled module requires the CSP/370 Runtime Services product at execution to function, so the CSP runtime environment must be defined to IMS. CSP/AE and the runtime environment are discussed more completely in later chapters.

APPLICATION DESIGN BASICS

Application design is something that has gotten much lip service over the years, but, sadly, it usually does not get the full attention it deserves. You will find that with CSP (as well as with most high-level tools), the assumption is that you have done a complete job of design before beginning development. If you have designed well, you will realize the full benefits of CSP's application development. If the application design is incomplete, much of the advantage of a product like CSP is forfeited. Time will be spent redoing and attempting to retrofit into already existing routines. This is where a product like CSP is at a disadvantage compared to older programming tools. (Older tools like COBOL and PL/I permit more flexible "tinkering" with their logic and execution paths. Unfortunately this often leads to a morass of less maintainable code.)

Many of the activities in a CSP application are based upon the I/O activities to be performed. If the design is not complete, then, "patching" (a time-honored tradition among 3GL programmers) is required. It may actually take longer to make major "patches" to a CSP application than it would take to completely rewrite the sections involved. This is because each process in CSP represents a logical function to be performed by the application. Thus, the application design is incorporated intimately into the application. This makes it more difficult to "fly by the seat of your pants" while creating an application. The product is designed to help those who have a plan. If you don't have a plan, you get what you deserve.

No matter what type of application design process is used, the basic premise of most is to identify the functional units—that is, to create a design that identifies the inputs and outputs of a system as well as the data transformation processes that are required in between. Any number of methods and documentation tools works well for this process, but one design documentation tool seems particularly suited to the design of CSP applications. The structure chart identifies individual blocks of an application and the relationship between routines. A structure chart does *not* describe any form of logic or control.

A sample structure chart is shown in Figure 1.1. Please note that the relationship between modules is depicted naturally. Processes that call other modules are higher on the chart. Subordinate processes are placed underneath those that use them. Processes that are reused are usually marked in one of two fash-

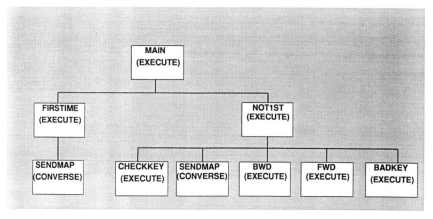

Figure 1.1.   Structure chart.

ions: a double line around the process or a slash across one corner of the process's box to depict its common nature. CSP applications fit the structure chart methodology since each CSP process can be made to represent a single "box" in the structure chart.

No matter what form of analysis and design methodology you use, and no matter what tools you use to depict that design (brackets, bubbles, lines, slings, arrows, outrageous fortunes) we all tend to go through the same basic steps.

To begin the design process, identify the problem that is to be solved in as simple a form as possible: what input data is available; what output(s) are desired; and what transformation process(es) are required to get from input to output successfully? Each type of I-O (input or output) activity should be represented by an individual "box" in the structure chart. In addition, each major transformation of the data should be represented, too. Though not required, it is a good idea to have a single "mainline" routine that causes all other activities to be performed. This provides a convenient place for control, and it also provides future maintenance personnel with a single location to find the overall logic path of the application.

The exercises presented here walk you through creation of various portions of a system to deal with the bookstore's inventory. The database that must be used includes entries for many technical tomes covering the gamut of the data processing industry as well as other exciting titles. Each book is represented by a record including the following:

Inventory ID
Title
Author(s)
Publisher
ISBN number
Library of Congress catalog information
Library of Congress catalog number
Dewey Decimal number
Cover type
Date published
Date last revised
Subject

Brief description
Number of pages
Number of illustrations
Replacement cost
Normal price
Sale price
Quantity on hand
Outstanding orders
Reorder level
Quantity on order

As new books are added to the inventory, the appropriate record is created/updated. If books are discontinued, the book's entry is deleted after the last sale.

A system is needed to report on the status of a given book and to allow updates, deletions, and additions of new books. The data now in the system was inherited from a previous inventory. Since some publications have no ISBN number, our system uses an internal number (inventory ID) to uniquely identify each book. These numbers are ever increasing and unique to each book.

STAGES IN CSP APPLICATION CREATION

This book will walk you through the creation of CSP applications a little at a time; no foreknowledge of CSP is assumed. Questions and exercises at the end of each chapter reinforce the topics in that chapter. In measured steps, you will complete all the activities necessary to create CSP objects and applications. The basic steps are similar to those in any programming environment:

1. Define the inputs/outputs.
2. Define the records.
3. Define the applications to use the inputs, outputs, and records.

CSP map definition allows us to "paint" the screen that the user will see. We'll also be able to control the color and display characteristics of the data. CSP even provides several built-in edit capabilities. Once defined, a map may be used by any number of applications without change.

CSP record definitions describe the data to be viewed or manipulated in the various file types supported by CSP and SQL rows. Both the record definitions and the data items definitions may be shared with future applications.

CSP applications are collections of routines called processes and statement groups, which provide the programming capability of CSP. As you might have guessed, these routines are reusable, too! CSP's pattern of reusable "objects" provides a sound foundation for future development if created as the result of a good design.

HOW THIS BOOK IS LAID OUT

An introduction to CSP is presented in Chapter 2, with the various components necessary for successful application creation. Chapter 3 covers creating a MAP (picture on a terminal), including the edit capabilities of CSP. In Chapter 4 you will learn how to create an application to display (and retrieve) a map and how to test it. The next step (Chapter 5) is learning the power of CSP's processing statements for data manipulation, calculations, conditional processing, and looping. Then Chapter 6 will show you how to define records so that CSP can use them. Chapter 7 goes into a CSP/AD utility function called the list processor. This is a very useful feature that will probably become one of your favorite places to start activities. In Chapter 8 you will see how data I-O is accomplished within CSP applications. The various types of data supported by CSP are explored along with basic read and write capabilities. CSP's built-in message processing facility is discussed in detail in Chapter 9. You will learn how to create and use messages with this tool. Another useful CSP feature are the data tables, the subject of Chapter 10. You will learn how to define these tables, load them with data, and use them from an application. Chapter 11 investigates the "sequential" processing capabilities of CSP. The mechanics involved and design considerations raised by serial processing of data are covered. We'll get into SQL processing in Chapter 12. You'll be exposed (yikes!) to the definition of SQL row records, the ability to modify CSP's SQL, cursors, and error processing. Transferring control from one application to another is the subject of Chapter 13. Each of the mechanisms supplied by CSP is covered along with a discussion about when and how to use them. Chapter 14 shows you how to use CSP's built-in utility program for manipulating MSL members and generating applications. Chapter 15 deals with using the External Source Format utility for "cloning" applications and transferring definitions to and from a Programmable Workstation (PWS). Another utility for manipulating ALFs is the subject of Chapter 16. TSO and the specifics of using CSP in that environment are the topics of Chapter 17, while CICS and its support for CSP are highlighted in Chapter 18. Finally, Chapter 19 gives a brief glimpse of what's ahead in CSP.

Just to give you a taste of what's to come, Figures 1.2 to 1.5 show a small CSP application. I know this looks like gibberish at this point, but I've found in training classes over the years that it's useful to give people a "sneak preview" of coming events.

Figure 1.2 shows the libraries used to hold the objects for this application. These libraries are called Member Specification Libraries (MSLs).

Figure 1.3 names the application, tells CSP what map group to use, and the name of a working storage record to be used.

Most applications use a "mainline" routine to control the application's logic. Figure 1.4 shows the statement that would be included in this application's main routine. Notice that the code

---

APPLICATION NAME: ZZ02APP   CSP/AD   DATE: 08/10/90 TIME: 11:36:36 PAGE: 00001

MSL CONCATENATION SEQUENCE

	MSL ACCESS ORDER	MSL FILE NAME
READ/WRITE MSL:	1	MYMSL
READ-ONLY MSL(S):	2	ROMSL
	3	ROMSL2
	4	
	5	
	6	

Figure 1.2.   MSL sequence.

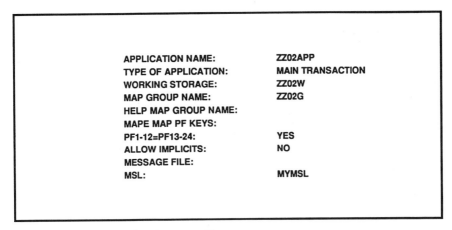

APPLICATION NAME:	**ZZ02APP**
TYPE OF APPLICATION:	**MAIN TRANSACTION**
WORKING STORAGE:	**ZZ02W**
MAP GROUP NAME:	**ZZ02G**
HELP MAP GROUP NAME:	
MAPE MAP PF KEYS:	
PF1-12=PF13-24:	**YES**
ALLOW IMPLICITS:	**NO**
MESSAGE FILE:	
MSL:	**MYMSL**

Figure 1.3.   Application specifics.

```
APPLICATION NAME: ZZ02APP    CSP/AD    DATE: 08/10/90 TIME: 11:36:36 PAGE: 00004

ZZ02MN  ***************************************************************************
        *                         MYMSL                                        *
        ***************************************************************************
        OPTION -  EXECUTE        OPTION

                  SET MAPEMP CLEAR;
                  SET EMP EMPTY;
                  WHILE EZEAID NOT PF3;
                    AND MAPEMP.QUITFLD NE 'QUIT';
                        PERFORM ZZ02CON;
                        IF EZEAID IS PF3;
                          OR MAPEMP.QUITFLD EQ 'QUIT';
                              EZECLOS;
                        END;
                        IF MAPEMP.EMPID NOT MODIFIED;
                              SET MAPEMP CLEAR;
                              MOVE 'PLEASE ENTER AN EMPNO' TO EZEMSG;
                        ELSE;
                              IF MAPEMP.EMPID IS MODIFIED;
                                    MOVE MAPEMP.EMPID TO EMP.EMPNO;
                                    SET MAPEMP CLEAR;
                                    PERFORM ZZ02INQ;
                              ELSE;
                                    PERFORM ZZ02CHG;
                              END;
                        END;
                  END;
```

Figure 1.4.   Mainline process.

```
APPLICATION NAME: ZZ02APP    CSP/AD    DATE: 08/10/90 TIME: 11:36:36 PAGE: 00010
RECORD DEFINITIONS


                         RECORD NAME:            ZZ02W

                         ORGANIZATION:           WORKING STORAGE
                         LENGTH IN BYTES:        49
                         DEFAULT SCOPE:          LOCAL
                         MSL:                    MYMSL

NAME        LVL  OCCURS  TYPE SCOPE LENGTH  DEC  BYTES START MSL DESCRIPTION
W-RECORD    05           CHA  LOCAL    49           49    1
W-CODE      10           CHA  LOCAL     3            3    1
W-SALARY    10           PACK LOCAL     9    2       5    4
W-QHAND     10           BIN  LOCAL     4            2    9
*           10           CHA  LOCAL    20           20   11
W-EMPNO     10           CHA  LOCAL     9            9   32
W-SSN       10           CHA  LOCAL     9            9   41
```

Figure 1.5.   Working storage.

listed is reasonably easy to understand, even though you might not be familiar with the syntax.

A vital part of definition is the use of record definitions. Figure 1.5 shows a Working Storage record's definition. Again, notice that even though the setting is unfamiliar, the basics of the record and data item definitions are simply presented.

AVAILABLE DOCUMENTATION

This book is intended to supplement the IBM documentation, not replace it. Find the repository of CSP manuals at your installation and make use of them. I've listed the IBM form numbers for the most important CSP manuals below:

Form Number	Manual
SH23-0502	CSP/AD Operation Development
SH23-0503	CSP/AD, CSP/AD System Administration

SH23-0505   CSP/AD, CSP/AD Message, Codes and Problem Determination

SH23-0515   CSP/AD Reference

SH20-6435   CSP/AD Developing Applications

SH20-6433   CSP/AD External Source Format Reference

SH20-6434   Defining Applications on the System/370

SH20-6765   Administering CSP/AD and CSP/AE on MVS

SH20-6766   Administering CSP/AD and CSP/AE on VM

SH20-6767   Administering CSP/AD and CSP/AE on VSE

SH20-6768   Administering CSP/AD and CSP/AE on DPPX

SH20-6769   Administering CSP/AD and CSP/AE on OS/2 and IBM DOS

SH20-0514   Generating and Running IMS Applications

2

Introduction to CSP/AD

This chapter covers the basics of CSP Application Development including definition of applications, screen formats, available function keys, and commands used. Some of these topics represent necessary "behind the scenes" knowledge and others represent basic information required to use CSP properly. The various stages of creating a CSP application involve the following:

1. creation (allocation) of Member Specification Libraries
2. application design
3. map definition
4. record definition
5. application and process definition

While it is possible to throw together an application in CSP, resist the temptation. Unlike other application programming tools you may have encountered, a design is essential for CSP application development. It may actually be easier to rewrite poorly designed CSP applications than to modify them. This chapter details the basics of CSP Application Development and lays the basic groundwork for the remainder of the book. If you already have some familiarity with CSP/AD, much of this chapter may be a review. However, if you are like most users of CSP, some of the "basics" covered here may be new to you as well.

CROSS SYSTEM PRODUCT "BUZZWORDOLOGY"

Throughout this text, several terms are used repeatedly when referring to CSP and CSP objects. This list is offered as a preview to define terms that occur frequently.

CSP Term	What It Means
CSP/AD	Cross System Product Application Development, where CSP applications are coded and tested
CSP/AE	Cross System Product Application Execution, the execution environment for CSP applications
SAA	System Application Architecture, IBM's standards for software systems development
CUA	Common User Access, part of SAA standards specifically relating to screen access
MSL	Member Specification Library, storage facility for CSP "source" code
ALF	Application Load File, storage facility for CSP "executable" code
Application	Program, set of activities to accomplish task
Process	Block of code, paragraph, procedure to accomplish a single application function
Map	Layout for screen or part of screen
Map group	Collection of maps associated with one application
Record	Definition of the result from one output operation (access method, name of file/database, key(s) needed, field definitions, and documentation)
Item	Field defined on map or in record
Table	Special CSP construct for predefined arrays

MEMBER SPECIFICATION LIBRARIES (MSLs)

Member Specification Libraries (MSLs) are special files created to store CSP definitions. In the MVS and VSE (IBM mainframe computer) worlds, MSLs are defined as VSAM Key-Sequenced Data Sets (KSDSs). The VSAM KSDSs used for MSLs are very volatile, that is, items are constantly being added, deleted, and updated. A VSAM KSDS works best when originally configured with this volatility in mind and when reorganized on a frequent basis to minimize any inefficiencies created by file gyrations. Care should be taken when setting the CA and CI freespace percentages and default buffer sizes for an MSL's KSDS. Traditional VSAM tuning methods work well for MSLs. The JCL and IDCAMS statements necessary to create an MSL are located in Appendix A of this book.

MSLs can effectively store very large numbers of definitions. However, it is probably best to break up definitions into several libraries based upon function and overall usage. Each individual application developer (programmer) should have his or her own MSL. Development/maintenance groups should have multiple libraries that are shared for read-only purposes and updated only in a controlled fashion. At a minimum, these should include a library of definitions currently in use in the production system, a library of common definitions to be "cloned," and a library for work in progress. Some organizations like to add special libraries for transferring definitions during the production-acceptance process. By using CSP's ability to concatenate libraries, developers may work on some definitions in their individual read-write libraries while sharing existing definitions from the read-only libraries. It is a good idea for application developers to purge items from their personal libraries as soon as they are complete and have been transferred to the group's shared libraries. Likewise, it is a good idea to purge items from work-in-progress libraries once they have been migrated to more permanent status. For safety's sake, don't forget that MSLs are just VSAM KSDSs and may be easily backed up using the IDCAMS REPRO utility. Though your shop probably has some automatic backup and recovery mechanism, you might consider making periodic "insurance" copies of your MSLs.

One negative feature of MSLs is that they must be edited using the CSP/AD product. Conventional source code editors do not provide access to VSAM KSDS, and those that allow access to VSAM data are unaware of the formatting that CSP requires. One exception to this is when a CSP object has been exported in External Source Format using the CSP/AD Utility. External Source Format is a highly specialized "tag" language used for communicating between CSP and other environments. External Source Format code is in a simple, flat-file, 80-byte record form easily modified by most editors. The syntax of External Source Format makes it impractical for original creation of applications, but it lends itself well to "cloning" existing code. Chapter 15 of this book is devoted entirely to the use of External Source Format and CSP/AD.

CSP objects are stored in MSLs with unique names. Since a KSDS allows access to data sequentially or by some key value, each CSP object's name becomes part of a unique record key in the VSAM KSDS. To minimize confusion and reduce effort, it is crucial to decide upon a naming convention early in the process of creating CSP applications. MSLs provide a powerful data dictionary-like capability that can be best utilized by following a naming standard.

MSL Object Naming Conventions

Given some of CSP's naming limitations (discussed later), IBM has suggested (and many shops follow) this standard.

1. First two characters denote a system or subsystem.
2. Third and fourth characters denote application within system.
3. Fifth character denotes type of object (map, process, and so forth).
4. Sixth and seventh characters are object number within type.
5. Additional characters are used for documentation.

For instance, a payroll system might use the first two characters "PY" and the first application in the payroll system would be numbered "01." Thus, "PY01" would serve as the prefix for all objects having to do with the payroll system's first application. It is also common to use a series of application IDs to denote commonly available objects (for instance, PY00 . . . for common payroll objects). This is the naming convention followed throughout this text.

After applications and objects have been defined in MSLs, they may be tested using the CSP/AD test facility (in most environments). For production-style execution it is necessary to generate definitions using the CSP/AD utilities and store the resulting output in Application Load Files (ALFs).

APPLICATION LOAD FILES (ALFs)

Application Load Files (ALFs) are libraries used to hold portions of CSP applications for execution under CSP/AE. ALFs are also used as repositories when transferring MSL members from one MSL to another. Like MSLs, ALFs are also represented by VSAM KSDSs in IBM mainframe environments. Once again, basic VSAM tuning techniques should be applied when defining and using ALFs. ALFs are not as volatile as MSLs and are accessed frequently in a skip-sequential manner.

After a CSP object (application, map, or table) has been tested under CSP/AD, it is important to generate it for use in the executable environment (CSP/AE). The generation process performs a syntax check and parsing of CSP's code. The result from the generation process is one or more members in an ALF. It is important to realize that ALF members do not represent fully executable code and must be supplemented by run-time routines in order to execute. CSP generation targets particular execution environments, and the ALF members resulting from the generation may be transferred to those environments freely. It is not uncommon for organizations to create special ALFs just for the transfer process. This is especially useful in environments like CICS where sharing of the ALF outside of CICS may lead to data integrity problems.

For those executing in MVS, sample JCL to create an ALF is listed in Appendix A of this text. However, MVS users should be aware that upon the release of CSP/AD Version 4 (fall 1992) CSP/AD will offer generation of VS COBOL II code for all MVS environments. This means that execution will be achievable using standard load modules and CSP Runtime Services. With CSP/AD Version 4, ALFs will be needed under MVS only for transferring objects or containing generated objects targeted for non-MVS environments.

Now that the basic library structures have been described, let's get into the CSP Application Development (CSP/AD) product.

GETTING INTO CSP/AD

In most installations, development occurs under the ISPF dialog manager. Entering CSP/AD is usually performed in one of two ways: via a menu selection choice or by using the command procedure (CLIST, EXEC, and so forth) "XSPD." If your installation chooses to use a menu selection choice, it might not be listed on the main ISPF menu. Look for submenus in ISPF with names like "AD/CYCLE," "Development Tools," "DB/2," or "Programmer Aids." You might even consider asking the systems programmers in charge of CSP in your installation how to gain access. Accessing CSP may require using special logon procedures in order to allocate the appropriate files and libraries. Another, less frequently used option is to execute the XSPD command procedure. This may be done from ISPF Option 6, or by issuing the command from any ISPF command line as follows:

```
( ==> TSO XSPD )
```

Under VM, the XSPD exec is executed to enter CSP/AD. The procedure to set up CSP/AD execution under DPPX/SP requires several steps. Refer to the DPPX/SP Operation manual for CSP. Under OS/2, execution of CSP/AD requires purchase of a separate product, the CSP Programmable Workstation (PWS). CSP's PWS product uses a Graphic User Interface (GUI) to simplify the CSP/AD process. Under the current release, objects defined using the PWS must be transported to another CSP/AD environment for testing and generation (External Source Format is used to transport the definitions). AS/400, IMS/TM, and MS-DOS do not support CSP/AD.

Some installations have added "front-end" panels prior to CSP/AD; others take you directly into the product. Most front-end panels provide places to define the MSLs, ALFs, and other files to be used during the CSP/AD session. Your local CSP administrator should be able to explain the use of any front-end panels in your system. CSP/AD's "main menu," the Facility Selection panel, is illustrated in Figure 2.1.

```
EZEM00        CROSS SYSTEM PRODUCT/APPLICATION DEVELOPMENT

==>
                        ENTER = Continue  PF3 = Exit
R/W MSL => yourmsl                                          HIGHEST MSL# =>1
.................................... FACILITY SELECTION ............................................


                   Enter number of facility desired => 7

                        1    List Processor
                        2    Definition
                        3    Test
                        4    Generation
                        5    Utilities and File Maintenance
                        6    Tutorial
                        7    MSL Selection

                        In all facilities:
                        PF1 = Help  PA2 = Cancel
```

Figure 2.1.   Facility selection panel.

```
EZEM01                   CROSS SYSTEM PRODUCT

==>
                 ENTER = Continue      PF3 = File and Exit

.................................... MSL SELECTION ....................................

          MSL FILE NAME        MSL ACCESS           CONNECT
                                 ORDER              STATUS

Read/Write MSL:
   => yourmsl                 => 1

Read-Only MSLs:
   =>                         => 2
   =>                         => 3
   =>                         => 4
   =>                         => 5
   =>                         => 6


                    In all facilities:
                 PF1 = Help      PA2 = Cancel
```

Figure 2.2.   MSL selection panel.

From this panel, selections are made to use different CSP/AD facilities for application creation, modification, and testing. Should your MSLs be unallocated or unavailable, CSP/AD's MSL selection panel is displayed as illustrated in Figure 2.2.

This panel will appear at CSP/AD initiation if any MSL defined in the user profile is unavailable or does not exist. If files have been allocated correctly and are available, make sure that the names have been spelled correctly. A problem might also arise if you have not been granted adequate security clearance to reach the MSLs in your list. If you are having problems, check with your CSP administrator.

TYPICAL CSP/AD SCREEN FORMATS

CSP/AD uses a fairly standard set of screens throughout the Application Development process. Unfortunately, these screens do not yet conform to the CUA standard, nor do they mimic the ISPF/PDF screens most developers are comfortable with. Instead, CSP/AD uses a standard screen as depicted in Figure 2.3.

```
EZEM29                          MAP DEFINITION
EZE00254A  This is a new map, you must choose at least one device
==>
                    PF3 = Exit (or continue if new definition)
                         Map Name = grpnam mapnam
...................................... DEVICE SELECTION ...............................................


                    Replace the * before the device name with:
                    A (Add), D (Delete), or R (Replace)
Total lines 28 ...............................................................................  Lines 16 to 28
         DEVICE NAME  LINES COLUMNS   DEVICE CLASS
    *      8775-3C      032    080     DISPLAY        MAY BE ADDED
    *      8775-4C      043    080     DISPLAY        MAY BE ADDED
    *      ANY-1D       012    080     DISPLAY        MAY BE ADDED
    *      ANY-2D       024    080     DISPLAY        MAY BE ADDED
    *      ANY-2D       032    080     DISPLAY        MAY BE ADDED
    *      ANY-3D       032    080     DISPLAY        MAY BE ADDED
    *      ANY-4D       043    080     DISPLAY        MAY BE ADDED
    *      ANY-5D       027    132     DISPLAY        MAY BE ADDED
    *      ANY-D        255    160     DISPLAY        MAY BE ADDED
    *      PRINTER      255    132     PRINTER        MAY BE ADDED
    *      PRINT-B      255    132     PRINTER        MAY BE ADDED
    *      3767         255    132     PRINTER        MAY BE ADDED
    *      5550D        024    080     DISPLAY(DBCS)   MAY BE ADDED
```

Figure 2.3. Sample screen layout.

line 1	Usually a title, depending on MAPID setting, the current MAP name will appear in the upper-left corner
line 2	Messages
line 3	Command entry (CSP commands only)
line 4	Function keys active on this screen
lines 5–7	Further information about the panel or its function (often a title)
lines 8–24	Rest of screen

A typical CSP/AD screen displays a title on the first line. If the MAPID ON command has been issued, the current map's name will display at the far left side of the first line. On some panels the far right side of the line displays "more" arrows indicating that scrolling keys may be used:

->   More to right
<-   More to left
<-> More both right and left

Line two shows any messages displayed by CSP/AD. Pressing PF1 (or issuing the HELP command manually) when a message is displayed will cause an explanation for the current message to display. For information on messages, see the "CSP/AD and CSP/ AE Messages, Codes, and Problem Determination" manual.

The third line allows entry of CSP/AD commands, which are covered later in this chapter. For TSO/ISPF/PDF users, it is important to realize that even though CSP/AD is an ISPF application, TSO and ISPF commands may not be entered from the CSP/ AD command line.

The fourth line (and sometimes the fifth) displays the special function keys available from the panel. They are listed according to the key used and followed by the key's function. If you are using a personal computer and are emulating a terminal, make sure that you have obtained a correct keyboard map. It is essential to know which personal computer keys correspond to desired terminal keys.

The fifth line is usually a subtitle describing specific objectives of the current panel. The sixth and subsequent lines contain data to be viewed, entered, or modified by the application developer. On some panels, the lower portion of the screen will provide an "entry area" with a line number area on the left allowing entry of line commands (covered later in this chapter). When line numbers are displayed, the panel is usually scrollable using CSP/AD function keys or commands.

Function Keys

CSP/AD has several predefined function keys. Each CSP/AD key is assigned a command that may also be entered from the command line. Several of the keys are universal in application, some have different meanings in different contexts, and some only work in a particular context. The chart below lists the keys used on IBM 327x-type terminals (also 318x, 319x, and 329x).

327x Function Key	Function and Description
PF1/PF13	HELP—Displays information about the current screen or message. If displaying information about a message, pressing PF1 again displays help for the current panel.
PF2/PF14	MOVE CURSOR—Swaps the cursor between the fixed area at the top of the display and the scrollable area at the bottom. This function is usually accomplished more easily using keyboard cursor control keys (tab, return, arrows, and so forth).
PF3/PF15	EXIT—Saves changes on the current panel and returns to the previous panel. When the EXIT key is pressed from within a multistep process, EXIT saves and displays the next panel in the process.
PF4/PF16	This button has different functions depending upon the function being used at the time. During Map Presentation this button issues TAB LEFT, moving the cursor to the left one tab stop (many developers find that CSP/AD's tabs

are not worth the effort to use them).

When defining an SQL row, PF4 causes the row's field definitions to be compared to the actual database definition. CSP accesses the database catalog and highlights the database columns represented incorrectly or unused. When viewing the application process list the overall structure of the application is displayed/refreshed.

PF4 causes expansion of %GET code when used during process definition, resets the SSA list when used during DL/I object selection, and controls the processing of an application test.

PF5/PF17	This key has two functions depending upon where it is used. During Map Presentation this button issues TAB RIGHT, moving the cursor to the right one tab stop (many developers find that CSP/AD's tabs are not worth the effort to use them). When testing and stopped at a break point, this key allows single-step testing. The test tool executes only one CSP command each time the key is pressed.
PF6/PF18	Should not be used from CSP/AD. This key simulates a CLEAR key and under TSO this may cause unpredictable results.
PF7/PF19	SCROLL BACK—Scrolls toward the beginning of the currently displayed data. Scroll amount is controlled by values entered with the SCROLL command. Some screens default to scrolling half a page, others a full page.
PF8/PF20	SCROLL FORWARD—Scrolls towards the end of the currently displayed data. Scroll amount is controlled by values entered with the SCROLL command. Some screens default to scrolling half a page, others a full page.
PF9/PF21	REMOVE or RESTORE—Toggles display of CSP/AD information on the screen. Allows display of full page for data entry.

"COMMAND LINE" COMMANDS

CSP/AD provides several commands that may be issued to control processing. These commands may be entered on the command line of most displays following the displayed arrow (==>) on the screen's second line. Several of these commands have meaning only in particular screens. Placing an ampersand (&) before any command causes it to remain in the command area for repeated execution. As an introduction, the command line commands are listed in three groups:

1. commands used on all displays
2. commands used during map definition/modification only
3. special-use commands

Syntax for these and other CSP commands is listed in Appendix A.

Commands Used on All Displays

Command	Function
?	Retrieve previous command(s) (up to 7)
+nnn	Scroll towards end of data nnn lines
−nnn	Scroll towards beginning of data nnn lines
ALARM n	Causes alarm to sound if n seconds elapses between user inputs
BOTTOM	Scroll to bottom of current list/data
CANCEL	End current function, ignore changes (same as PA2 key)
CHANGE/CHG	Alter text strings as directed (more later)
EXIT	Leave current function and save changes (same as PF3/PF15)
FIND	Find or Locate text string (more later)
LOCATE	Same as FIND
MAPID	Toggles display of CSP/AD map ID in upper left corner of each screen

MESSAGE	Displays the requested CSP/AD message (see message file utility later in this book)
QUIT	Return to CSP/AD Facility Selection or List Processor display, leave current function, and ignore changes (same as repeated use of CANCEL or PA2)
RETURN	Save and return to CSP/AD Facility Selection or List Processor display (same as repeated use of PF3)
SCROLL	Set scroll values (see SCROLLING later)
TOP	Scroll to first line

Basic Commands Only for Map Displays

The following commands have value when defining maps. The section on map definition includes more detail on these commands.

Command	Function
ATTRIBUTE	Changes attributes for current field
CODES	Change or display Constant, Variable, and Spacer definition codes for map presentation
COPY	Copy lines defined by FROM-TO AFter or BEfore the current line
DELETE	Deletes specified number of lines (includes current)
FROM	Marks the beginning of a block to be used in a BOX, COPY, MOVE, REPEAT, or SHIFT command
INPUT	Insert lines into map AFter or BEfore the current line
MOVE	Move lines defined by FROM-TO AFter or BEfore the current line
REPEAT	Repeat lines defined by FROM-TO the specified number of times

SAVE	Save changes, do not leave display
TEST	Display map with specified fill characters in uninitialized protected fields
TO	Mark the end of a block to be used in a BOX, COPY, MOVE, REPEAT, or SHIFT command

Other "Command Line" Commands

These commands are listed for completeness. The MOVE and SHOW commands will be explained more thoroughly in the section describing the CSP TEST facility.

Command	Function
COPYLIST	Copy all members on the current list from read-only MSLs to the read/write MSL
MOVE	Change storage during testing (see TEST FACILITY later in this book)
RESET	Ignore and remove "line" commands, remove COLUMN and MASK settings
SHOW	Display storage during test (see TEST FACILITY)

Find "Command Line" Command

See Figure 2.4.

FIND, LOCATE, and /	may be used synonymously
FIND	may be abbreviated to F, LOCATE to L
textstrg	contains string to be found (located)
First	find the first occurrence of textstrg
Last	find the last occurrence of textstrg
Next	find the next occurrence of textstrg following the current cursor position
Prev	find the closest occurrence of textstrg in front of the current cursor position
ASis	requires exact match on case

	Find	textstrg	First or Last
or	Locate		or Next
or	/		or Prev
			ASis

Figure 2.4.   Find command syntax.

Several rules apply to the coding of textstrg:

1. Single quotes/apostrophes are used as delimiters by CSP. (Quotation marks (") are seen only as data.)
2. If the string contains embedded blanks or special characters, single quotes/apostrophes (') must be used to delimit the string ('toy dog').
3. For a single quote, use two single quotes together ('fred''s toy dog').
4. For maps, string may not contain characters defined using CODES.
5. Strings of unequal lengths are padded with blanks for comparisons.

Examples of Find Command

FIND BOB	finds next occurrence of BOB, bob, or Bob (case does not matter)
f 'aardvark'	finds next occurrence of AARDVARK, aardvark, or Aardvark (once again, case does not matter)
f 'Fred''s toy dog' n as	finds next occurrence of Fred's toy dog following the current cursor position (case match is required by AS option)
/ Koala AS	finds next occurrence of Koala (note that KOALA and koala would not be found since AS requires a match on case)
l "Smith" first	finds the first occurrence of "Smith" or "SMITH" or "smith" (case matches are not required since AS was not specified). Note that the double quote (") is seen as DATA by CSP, not as a delimiter.

Change	**oldtext**	**newtext**	**All or First or Last or Next or Prev or Rest ASis**

Figure 2.5.   Change command syntax.

Change "Command Line" Command

See Figure 2.5.

CHANGE   may be abbreviated to CHG or C

(*Note:* CHANGE does not work on VARIABLE FIELD NAMING screen.)

oldtext	contains string to be changed
newtext	contains replacement string for oldtext
All	change every occurrence of oldtext to newtext
First	change only the first occurrence of oldtext
Last	change only the last occurrence of oldtext
Next	change the next occurrence of oldtext following the current cursor position
Prev	change the closest occurrence of oldtext in front of the current cursor position
Rest	change all remaining occurrences of oldtext following cursor
ASis	requires exact match on case for search string

For both oldtext and newtext several rules apply:

1. If the string contains embedded blanks or special characters, single quotes (') must be used to delimit the string ('toy dog').
2. For a single quote, use two single quotes together ('fred''s toy dog').
3. For maps, string may not use characters defined as CODES.
4. Strings of unequal lengths are padded with blanks for comparisons.

Examples of Change Command

CHANGE BOB TED ALL

changes all occurrences of BOB, bob, or Bob to TED; like FIND, case is unimportant on the search argument unless AS is specified, however, the result's case is not changed.

c 'aardvark' 'kangaroo' a

changes all occurrences of aardvark to kangaroo (any combination of upper case and lower case in AaRdVaRk will match since AS was not specified, but the result would be lower-case kangaroo regardless).

c 'Fred''s toy dog' 'Zorro''s toy dog' r

changes all occurrences of Fred's toy dog following the current cursor position to Zorro's toy dog (case unimportant in search value).

CHG Koala 'QUANTAS mascot' A AS

changes all occurrences of Koala to QUANTAS mascot (note that KOALA would not be changed since AS requires a match on case).

CHANGE 'COBOL' 'CSP' first

changes the first occurrence of COBOL to CSP (once again, the command would be case-insensitive regarding the search string CoBoL).

Scrolling

On many CSP edit screens, it is possible to SCROLL using the designated function keys:

PF7/PF19	SCROLL UP
PF8/PF20	SCROLL DOWN
PF10/PF22	SCROLL LEFT
PF11/PF23	SCROLL RIGHT

| **SCroll** | **Left(x)** | **Right(x)** | **Forward(x)** | **Back(x)** |

Figure 2.6.   Scroll command syntax.

The amount scrolled may be controlled by using the SCROLL command to set the default number of rows/columns in each direction. All scrolling is done in units relative to the displayed edit area (see Figure 2.6).

Scroll amounts "x" may be set to the following values:

PAGE or P   Scroll one full page in the edit area

HALF of H   Scroll one half page in the edit area

MAX or M   Scroll all the way to the top/bottom/left/right

nnn   Scroll nnn lines/columns

Changes made to scroll values are temporary and are reset to the defaults when you exit CSP/AD.

After issuing the command in Figure 2.7, PF7/PF19 and PF8/PF20 scroll up or down only one half page.

| **SC B(HALF) F(HALF)** |

Figure 2.7.   Scroll command example.

Other COMMAND LINE commands used for SCROLLING include the following:

+nnn   Scroll nnn lines toward the end of the data

−nnn   Scroll nnn lines toward the beginning of the data

BOTtom   Scroll all of the way to the end of the data

TOP   Scroll all of the way to the beginning of the data

CSP "LINE COMMAND" COMMANDS

During CSP Record and Process definition, edit areas are available for entry by the application developer. The first column of these edit areas is represented usually by a "line number area" containing either a relative line number (0nn) or by the following symbol (−−>). This line number area may be overtyped with several commands. The line commands are very similar to those used by the ISPF/PDF product and some used by XEDIT. However, a significant difference is that CSP requires the line commands to be left-justified in the line number area.

Line commands affect the line they are entered on and sometimes other lines depending upon the options specified. Commands followed by a number indicate that the command is to act upon the specified number of lines beginning with the line the command is entered on. Commands using double characters (CC, DD, >>) impact a block of lines. When issuing block commands, the double character is entered both on the line that begins the block and on the line that ends the block.

Line Commands Used for Most Edit Areas

Line CMD	Function
/	Locate, shifts current line to top of display
COL	Show ruler line (not part of data)
MSK	Show/update mask line
A	Copy/Move lines After this line
B	Copy/Move lines Before this line
C	Marks line to copy
Cn	Marks n lines to be copied (beginning with this one)
CC	Marks block of lines to be copied
D	Delete this line
Dn	Delete n lines (beginning with this one)
DD	Mark block of lines for deletion

I	Insert a new line
In	Insert n new lines
M	Marks line to move
Mn	Marks n lines to be moved (beginning with this one)
MM	Marks a block of lines to be moved
R	Repeat this line
Rn	Repeat this line n times
RR	Repeat block of lines once
RRn	Repeat block of lines n times

Line Commands Used Only in Prologue and Process Definition

>	Shift current line right 1 position
>n	Shift current line right n positions
>>	Shift current block of lines right 1 position
>>n	Shift current block of lines right n positions
<	Shift current line left 1 position
<n	Shift current line left n positions
<<	Shift current block of lines left 1 position
<<n	Shift current block of lines left n positions

CSP HELP FACILITY

CSP/AD provides an extensive, screen specific HELP facility that may be used from any screen. At any time while in CSP/AD, press the PF1 (PF13) key or enter the HELP command to view information about the current screen. To explain an error message, press the HELP key while the message is displayed. Pressing HELP a second time after reviewing a message's HELP displays the current screen's HELP. The CSP HELP facility may also be used when defining CSP applications. Whenever the HELP key is pressed while executing a user CSP application, CSP automatically searches for and displays the appropriate HELP screen (if defined).

REVIEW OF GENERAL SCREEN PROCESSING

Throughout CSP/AD certain navigational capabilities are present on every screen. They are listed here for your convenience.

- To save changes and exit a function press PF3 or PF15 (PF3/ PF15 may be used repetitively to "back out" of a series of screens).
- To save changes and return to the CSP/AD Facility Selection display, key in the RETURN command.
- To ignore changes and exit the current function, press PA2 (PA2 may be used repetitively to "back out" of a series of screens.)
- To ignore changes and return to the CSP/AD Facility Selection display, key in the QUIT command.
- The PF1 key or HELP command may be used on any panel.
- =M (direct branch to Main Menu) may be keyed in on any command line to return to the CSP/AD Facility Selection display. (See Facility Transfer commands later in the text for more details.)

CHAPTER 2 EXERCISES

Questions

You should be able to answer the following questions:

1. Who is your CSP administrator?
2. What is an MSL used for?
3. Who defines MSLs in your installation?
4. Which Read-Only MSLs have been made available to you?
5. What are ALFs used for?
6. Who defines ALFs in your installation?
7. Which ALFs have been made available to you?

8. What environment does your installation execute CSP/AD under?
9. What environment does your installation execute CSP/AE under?
10. Do you use standard IBM 3270 keyboards? If not, do you have a copy of the appropriate keyboard chart?

Lab Setup for Future Exercises

1. Create a read-write MSL to use for completing the exercises included in this text. Request a small amount of storage (one cylinder or less) for the file.
2. Create an ALF to use for completing the exercises included in this text. Request a small amount of storage (one cylinder or less) for the file.

Map Definition

CSP uses the term MAP to define the data that appears on a terminal screen. MAPs usually define the entire screen as it appears to the application's client, but it is possible to define MAPs as portions of a screen. CSP groups MAPs together for storage and reference purposes, thus the term used is MAP GROUP. MAP GROUPs may contain many MAPs. Each CSP application is defined to use a single MAP GROUP. In addition, a HELP map group may be assigned to each application.

To define a map, it is generally a good idea to have planned the format of the map beforehand. Using paper and pencil, draw a picture of the map as you envision it (327x users, don't forget to leave room for an attribute byte before every field). Decide what data fields will be displayed, enterable, or changeable. Write down the edit criteria for each field. Finally, determine which types of terminals will be used to test and execute the application. Now you are ready to define the map to CSP.

From the main Facility Selection menu, select option 2 (Definition) to begin the map definition process (see Figure 3.1). The next panel you see will be the DEFINITION SELECTION panel (similar to the one following.) Enter the map group name, followed by a blank space, followed by the map name. Map group

```
EZEM00          CROSS SYSTEM PRODUCT/APPLICATION DEVELOPMENT

==>
                    ENTER = Continue      PF3 = Exit
R/W MSL = yourmsl                                        HIGHEST MSL# = 1
...............................................FACILITY SELECTION...............................................



            Enter number of facility desired => 2

                1    List Processor
                2    Definition
                3    Test
                4    Generation
                5    Utilities and File Maintenance
                6    Tutorial
                7    MSL Selection

            In all facilities:
            PF1 = Help        PA2 = Cancel
        5668-813 (C) COPYRIGHT IBM CORPORATION 1980, 1990
```

Figure 3.1.   Facility selection menu.

names may not exceed six (6) characters, while map names may not exceed eight (8) characters. Names may not include embedded blanks and may not begin with the characters "EZE." Key in member type 3 and press ENTER to continue the map definition (see Figure 3.2).

DEVICE SELECTION

If you are creating a new map, the Device Selection screen should appear next (see Figure 3.3). Many devices are listed for your selection. The device currently being used by CSP/AD will be highlighted. Place a character "A" before each type of device this map might appear upon. While there are some generic device types, it is best to choose those types applicable to your environment. Choosing specific device types will result in smaller modules and faster run times. If you accidentally select a device, delete it or replace it using the codes provided. Press PF3 (not ENTER) to continue the map definition process.

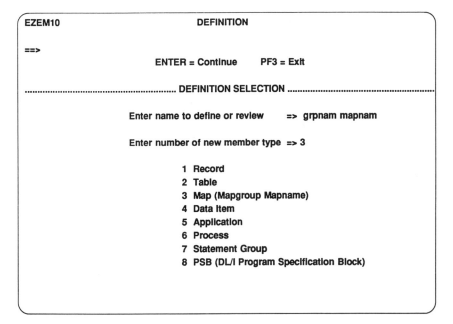

```
EZEM10                          DEFINITION

==>
                       ENTER = Continue     PF3 = Exit

.......................................... DEFINITION SELECTION ..................................................

              Enter name to define or review      => grpnam mapnam

              Enter number of new member type  => 3

                        1  Record
                        2  Table
                        3  Map (Mapgroup Mapname)
                        4  Data Item
                        5  Application
                        6  Process
                        7  Statement Group
                        8  PSB (DL/I Program Specification Block)
```

Figure 3.2.   Definition selection.

```
EZEM29                       MAP DEFINITION
EZE00254A  This is a new map, you must choose at least one device
==>
                     PF3 = Exit (or continue if new definition)
                     Map Name = grpnam mapnam
.................................................. DEVICE SELECTION ..........................................

              Replace the * before the device name with:
              A (Add), D (Delete), or R (Replace)
Total lines 28 ..................................................................................................... Lines 16 to 28
       DEVICE NAME   LINES COLUMNS   DEVICE CLASS
  *     8775-3C        032   080       DISPLAY        MAY BE ADDED
  *     8775-4C        043   080       DISPLAY        MAY BE ADDED
  *     ANY-1D         012   080       DISPLAY        MAY BE ADDED
  *     ANY-2D         024   080       DISPLAY        MAY BE ADDED
  *     ANY-2D         032   080       DISPLAY        MAY BE ADDED
  *     ANY-3D         032   080       DISPLAY        MAY BE ADDED
  *     ANY-4D         043   080       DISPLAY        MAY BE ADDED
  *     ANY-5D         027   132       DISPLAY        MAY BE ADDED
  *     ANY-D          255   160       DISPLAY        MAY BE ADDED
  *     PRINTER        255   132       PRINTER        MAY BE ADDED
  *     PRINT-B        255   132       PRINTER        MAY BE ADDED
  *     3767           255   132       PRINTER        MAY BE ADDED
  *     5550D          024   080       DISPLAY(DBCS)  MAY BE ADDED
```

Figure 3.3.   Device selection.

MAP SPECIFICATION

The next screen in the map definition process is the MAP SPECI-FICATION panel (see Figure 3.4). Here you need to provide CSP with specifics concerning the dimensions of the map and other details. Map size describes the overall height and width of the screen. It is usually best to choose 24 lines and 80 columns. Map position is used when the map is not to be displayed in the upper right corner of the screen. Having multiple small maps offers no performance benefit and adds a degree of complexity. It is probably best to limit your map designs to one per page starting in line 1 column 1.

CSP provides a certain amount of automatic editing for data fields. You may allow map users to skip these edits by naming up to five keys as BYPASS keys. Naming of bypass keys may be performed here or during application definition, though application definition is probably a more precise spot for this. Some terminals allow the entry of lower-case letters. CSP will automatically convert lower-case letters to upper case for all entries on this map when you code Fold Input = Yes. It is usually better to set this to

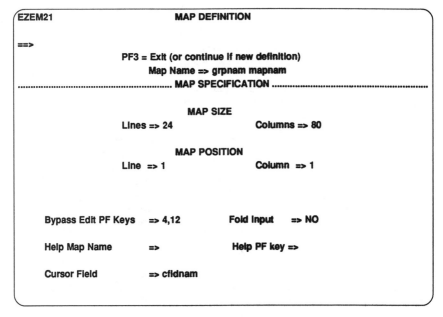

Figure 3.4. Map specification.

NO and specify upper-case translation on a field-by-field basis as desired.

Note: Not all terminals and controllers support lower-case letters. Make sure that the terminal, controller, and software (CICS TCT for example) are set to support lower case if it is needed for your task. CSP's built-in HELP system may be used by application developers. The name of a HELP map associated with the map being defined is keyed into the Help Map Name field and the number of the HELP key to be used is specified (or it defaults to PF 1). You must still define the HELP map and map group, then specify the HELP map group name for any applications that will use it. If named variables exist on the map, the Cursor Field will show the first unprotected field on the map or the cursor field if one is specified. If no named variables exist, then this field will not appear. Press PF3 to continue with the map definition.

MAP DEFINITION (SCREEN PAINTER)

The Map Definition panel is used to "paint" the screen; therefore, the nickname "screen painter" is often used. If you have worked with IBM's SDF/CICS or GDDM products, you should find this process very familiar. Since most CSP applications run on IBM 327x terminals, you should be aware that in front of every field the terminal reserves one character to record the field's ATTRIBUTEs. Attribute bytes contain indicators controlling a field's brightness, protection, and other settings. Attributes are discussed extensively later in this chapter.

See Figure 3.5. Two special characters are used to mark the beginning of fields. These characters are entered on the Map Definition panel wherever the attribute for a field will be. One character (CSP default is the pound sign "#") is used to define constants on the screen. Constants are on-screen text that will not change or be altered by the user, such as titles and column headings. Another character (CSP default is the not bar "¬"), is used to define variables on the screen. Variables are any screen item that will be entered/changed by the user or changed by the program. If you wish to change the attributes (intensity or protection) of a field, it must be defined as a variable. These two characters display near the top of the Map Definition panel with

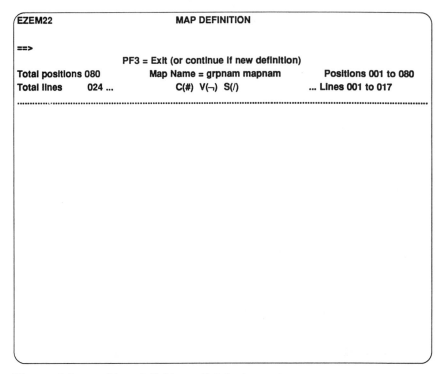

Figure 3.5.   Map definition—Original panel.

a third character, the spacer (CSP default is a slash "/"). The spacer character is useful only the very first time data is entered on a given line. The first time data is entered on a line, the spacer character is replaced by multiple blanks. Placing a spacer to the left of a field or group of fields causes it to be right-justified. To center a field or group of fields, key a spacer at the beginning and end of the field/group. Note that spacer characters are inserted based upon the current position of the cursor and the left/right bounds of the map being defined. In addition, the first time data is entered on a given line, a field size may be specified by following the variable character with a comma and a numeric value (no blanks). For instance, the string "¬,8" causes CSP to place a constant character (#) eight bytes after the variable character (¬), causing an eight-byte field to be defined. All enterable fields should be specified in this manner or by counting the field sizes yourself and following fields with a constant character. Figure 3.6 illustrates a sample screen BEFORE the developer has pressed the enter key.

After pressing the enter key, the screen looks like Figure 3.7. Notice the following automatic changes made by CSP:

- The first line is centered.
- Variables following QUANTITY ON HAND and QUANTITY ON ORDER have been expanded to nine characters each and ended by constant markers. Notice that the original space between the fields is preserved.
- Variables following MIN ORDER QTY and MAX ORDER QTY have been expanded, constant markers have been added, the original space between the fields is preserved, and the entire line (including constants) has been centered.
- Finally, the last line has been right-justified.

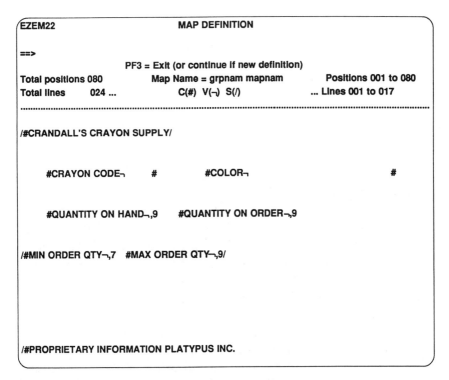

Figure 3.6.   Map definition—Special codes.

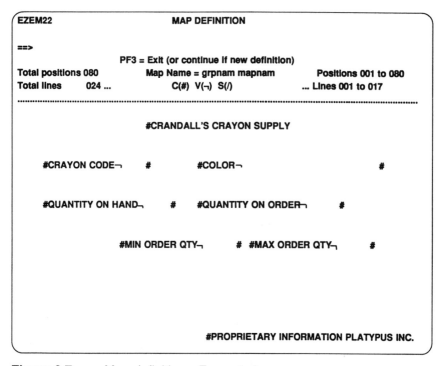

Figure 3.7. Map definition—Expanded.

The Constant, Variable, and Spacer characters may be altered using the CODES command. The CODES command is discussed later in this chapter.

Be careful! CSP's screen painter treats the entire screen as one long field if the map is defined as 24x80. This means that using the keyboard DELETE, INSERT, and ERASE EOF keys may yield undesired results. Some people define maps as 24x79 originally to avoid this, then alter the definition after screen painting is complete. This can cause problems later should you forget to change the map. It is probably easier to simply exercise caution while editing. The CSP screen painter is sometimes easy to confuse. Should you overkey an existing field and add a new field at the same time, the screen painter sometimes responds with an error message. Pressing PF9 (remove and restore CSP information) often solves this problem; otherwise, you may have to rekey the line. It is a good idea to issue the SAVE command before making major changes to the map.

Commands Only for Map Definition

Command	Function
ATTRIBUTE	View or alter a field's attributes
BOX	Draw box around FROM-TO area (IBM 5550's only)
CODES	View or alter constant, variable, or spacer codes
COPY	Copy lines defined by FROM-TO AFter or BEfore the current line
DELETE	Deletes specified number of lines (includes current)
FROM	Marks beginning of a block for BOX, COPY, MOVE, REPEAT, or SHIFT command
HEX	Places hex string at current location
INITIAL	Initializes DBCS fields
INPUT	Insert lines into map AFter or BEfore current line
LSHIFT	Shift block marked by FROM-TO left the number of columns specified
MOVE	Move lines defined by FROM-TO AFter or BEfore the current line
POSITION	Place the cursor in the position specified
REPEAT	Repeat lines marked by FROM-TO the specified number of times
RSHIFT	Shift block marked by FROM-TO right the number of columns specified
SAVE	Save changes, do not leave display
SET	Toggle FOLD/NOFOLD, FOLD causes entered data to be converted to upper-case
TABS	Set/display tabs during map presentation

TEST   Display map with specified fill characters (optional) in uninitialized protected fields

TO   Mark the end of a block for BOX, COPY, MOVE, REPEAT, or SHIFT command

Many of these commands are explained more fully on the following pages.

Changing Codes

Sometimes the characters chosen by CSP for Constants (#), Variables (¬), and Spacers (/) conflict with installation desires. The CODES command is used to change one or more of these characters. Issuing just the command CODES from the Map Definition panel results in a list of the three current CODES. See Figure 3.8.

CODES C(x) V(y) S(z)

Figure 3.8.   CODES command syntax.

Simply overkey the displayed values and press ENTER. Another option is to key in the command followed by the desired changes. For instance, CODES V(^) resets the variable identifier to "^" and leaves the other values unchanged. Be careful to use values that will not appear in the text constants appearing on the screen. DBCS codes may be changed also. See Figure 3.9.

CODES KCC(xa) KCV(xb) MCC(xc) MCV(xd)

Figure 3.9.   DBCS CODES command syntax.

From, To, & Current Line

When using COPY, MOVE, REPEAT, or xSHIFT, the FROM and TO commands are used to identify the top and bottom lines (or right and left sides for xSHIFT) of a block to be acted upon.

To mark a block beginning, do the following:

- Key in the word FROM on the command line.
- Place the cursor on the first line (leftmost position) of the area to be moved/copied/repeated/shifted and press ENTER.

To mark a block end, do the following:

- Key in the word TO on the command line.
- Place the cursor on the last line (rightmost position) of the area to be moved/copied/repeated/shifted and press ENTER.

To mark a single line, only FROM is required.

Many map definition commands depend upon the placement of the "current" line. To make a line current, place the cursor on that line and press the ENTER key.

Copy & Move

Any time lines are copied or moved, there MUST be enough room in the map (empty lines) to allow the copy/move. Either MOVE, INPUT, or DELETE lines to make room. The syntax for COPY follows. See Figure 3.10.

To copy a block, do the following:

- Make sure a block has been marked properly using FROM & TO.

COPy AFter   or   COPy BEfore

Figure 3.10.   COPY command syntax.

- Key COPy AFter or COPy BEfore on the command line (AFter is the default and may be omitted).
- Move cursor to the line you wish to copy the data AFter or BEfore.

The syntax for the MOVE statement is shown in Figure 3.11.

MOVe AFter   or   MOVe BEfore

Figure 3.11.   MOVE command syntax.

To move a block, do the following:

- Make sure a block has been marked properly using FROM & TO.
- Key MOVe AFter or MOVe BEfore on the command line.
- Move cursor to the line you wish to move the data AFter or BEfore.

Careful! MOVE will erase the data from its current location.

Repeating Lines

The syntax for the REPEAT command is simple—just follow the command with the number of repetitions desired (1 is the default). The system will repeat to cause the data to overflow the defined page size. It may be necessary to MOVE or DELETE lines to make room for the repetition. See Figure 3.12.

```
Repeat n
```

Figure 3.12.   REPEAT command syntax.

To repeat a block, do the following:

- Make sure block is marked properly using FROM & TO.
- Type R n to repeat the block n times (you may need to delete or move lines first to make room).

To repeat the current line, do the following:

- Do NOT specify FROM-TO.
- Type R n to repeat the current line n times.

RShift and LShift

Shifting data is also a matter of simply following the LSHIFT or RSHIFT command with the number of columns the data is to be shifted (again, 1 column is the default). CSP will not allow you to shift data off the page. See Figure 3.13.

LShift n	**or**	**RShift n**

Figure 3.13.   Syntax of shift commands.

To shift a block of lines left or right, do the following:

- Make sure a block is marked properly using FROM & TO.
- Type LShift n or RShift n to shift the block.

To shift the current line, do the following:

- Do NOT specify FROM-TO.
- Type LShift n or RShift n to shift line.

Again, you may only shift into empty columns. At least as many empty columns must be available to the right or left for the shift to work properly. If a line is unshiftable, the FROM value is set to that line by CSP.

Inserting Lines

Lines may be inserted into the map using the INPUT command. There must be at least as many blank lines at the end of the map as you are adding. It may be necessary to use the MOVE or DE-LETE commands to make room. As the syntax in Figure 3.14 illustrates, you may specify that the new lines are to be placed either before or after the current line (after is the default). CSP defaults to 1 line, but the number of lines must be specified if AFter or BEfore are entered. Again, new lines are added either AFter or BEfore the current line (default is AFter).

INPUT n	AFter or BEfore

Figure 3.14.   INPUT command syntax.

To add two lines, do the following:

- Type INPUT 2 on the command line.
- Move the cursor to the line where you wish to add the new lines after pressing ENTER.

Deleting Lines

Lines may be deleted using the DELETE command. DELETE is often used to make sure enough empty lines are at the end of a map to allow INPUT, COPY, or MOVE operations. Once more, the syntax is simple. Look at Figure 3.15. The command illustrated above deletes n lines (default is one line). Line deletion begins with the current line and deletes successive lines based upon the number entered.

DELETE n

Figure 3.15.   DELETE command syntax.

To delete two lines, do the following:

- Type DELETE 2 on the command line.
- Move the cursor to the first of the two lines to be deleted.
- Press ENTER.

Drawing Boxes on the Screen

Some newer types of devices (IBM 5550 terminals) allow the creation of outlines around data. CSP supports this capability by providing the BOX command (see Figure 3.16). To draw a box, do the following:

- Mark the area to be outlined using FROM & TO.
- Enter the command BOX.

Be careful! BOX creates single-character constants that may overlay other constant data on the screen. CSP will create the necessary constant field codes (#) when it adds the constants. FROM and TO must truly represent the upper left and lower right limits of the box or an error will result. BOXes may not overlay variable fields, DBCS fields, or MIX fields.

BOX

Figure 3.16. BOX command.

MAP DEFINITION COMMANDS

POSITION, HEX, and INITIAL are all commands used while defining maps, though their usefulness may be limited (see Figure 3.17). The POSITION command moves the cursor to the position specified. If both Line and Column are left out, the current cursor position is displayed. If only one of the two attributes is left out, the current cursor position is used in its place.

POSition Line(x) Column(y)

Figure 3.17. POSITION command syntax.

The HEX command places the hexadecimal value 'hexstring' into the map at the current cursor position (see Figure 3.18). Be careful not to use a combination of hex values that happens to be a terminal command (see the hardware's literature). It is not usually a good practice to place hex data on a screen. HEX may be used to place values into a DBCS or MIX field. Mixed fields may require the addition of the '0E' (shift out) and '0F' (shift in) characters as well.

HEX hexstring

Figure 3.18.   HEX command syntax.

DBCS INITIALIZATION

DBCS or MIXed fields (mixed DBCS and single characters) may be initialized using the command in Figure 3.19. This command is used to fill DBCS / MIX constant and variable fields during the Map Field Initialization display. KCC(xa) and KCV(xb) place the DBCS character 'xa' or 'xb' (respectively) into all uninitialized DBCS constant or variable fields (the default is '+'). MCC(xc) and MCV(xd) place the MIX character 'xc' or 'xd' (respectively) into uninitialized MIX constant or variable fields (the default is '¬').

INITial KCC(xa) KCV(xb) MCC(xc) MCV(xd)

Figure 3.19.   INITIAL command syntax.

TABS

Tabs may be set in the screen painter to ensure accurate positioning of columnar data. The TABs command is used to view the current tab settings or to create new tab settings. There are no default tabs. The specific syntax for the TABs command is seen in Figure 3.20.

TABs   col1 col2 col3 ...   coln

Figure 3.20.   TABS command syntax.

The TABs command specifies the column number(s) to be used as tabs. Once TABs are defined, the TAB LEFT function key (PF4/PF16) causes the cursor to move one tab stop to the left and the TAB RIGHT function key (PF5/PF17) causes the cursor to move one tab stop to the right. The keyboard tab keys are not used by the CSP TABs function, only the appropriate PF keys. Users of mainframe systems may find the response time less than desirable when using CSP TABs, because the tabbing commands are executed by the system and not by the terminal.

SAVING YOUR WORK

Issue the SAVE command (Figure 3.21) any time you wish to save the map definition. This is a good idea if you work in a "flakey" system or if you are about to make major changes. The SAVE command insulates you against inadvertent edit errors and system/terminal failure. SAVE makes all map changes permanent and cannot be undone by the PA2/CANCEL function.

SAVe

Figure 3.21.   SAVE command.

MAP TESTING

At any time while using the MAP PRESENTATION (painter) panel, you may enter the TEST command to display your map as the user will see it. You may then check to see that fields have been displayed and protected as designed. The TEST command provides a special feature allowing display of selected "fill" data in protected or auto-skip fields that are uninitialized. The command in Figure 3.22 causes the map to be displayed with exclamation marks filling all of the uninitialized protected/auto-skip

---

TEST FIL(!)

Figure 3.22.   TEST command syntax.

fields. You may key data into any unprotected variable fields. Unfortunately, the TEST command does not activate the built-in CSP edits (which are discussed later in this chapter).

One other command, ATTRIBUTE, which is used to view and change a field's attributes, will be discussed later in this chapter also. After painting the screen and testing it, press PF3 to continue the map's definition.

MAP DEFINITION—VARIABLE FIELD NAMING

If continuing a map definition from Map Definition panel (screen painter), the panel in Figure 3.23 should appear next. Each variable field on the map being defined is shown with a position number. Scrolling on this panel is unconventional. Change the "Number of first field to name" to get more name slots (or press PF9, position the cursor on the desired field, and press ENTER).

```
EZEM24                        MAP DEFINITION

==>
                      PF3 = Exit (or continue if new definition)
0001 <= Number of first field to name            Map Name = grpnam mapnam
.............................................. VARIABLE FIELD NAMING ..............................................
          NAME
        1 FNAM1
        2 FNAM2
        3 FNAM3
        4 FNAM4




Total positions 080                              Positions    001 to 080
Total lines   024 ................................................................. Lines   001 to 010

     CRAYON CODE 1            COLOR 2                         #


     QUANTITY ON HAND 3       QUANTITY ON ORDER  4
```

Figure 3.23. Map definition—Variable field naming.

Using the corresponding name area above, enter names for each field using the slots that correspond to the number shown in the desired field. Names must follow these rules:

1. There can be 1 to 32 characters (only 8 before CSP V3 R3).
2. The first character must be letter A–Z, $, #, or @.
3. The remaining characters must be A–Z, 0–9, $, #, @, –, or _ (– and _ are new with V3 R3).
4. There can be no embedded blanks.
5. The first three characters may not be "EZE" (except EZEMSG).

It is a good practice to have a message field defined at the bottom of every screen (or at the top if following CUA standards). Use EZEMSG to name your message field. This allows use of CSP's automatic message capability. Should you have an array on the map being defined, you must name every single representation of the field. Each field must be subscripted to illustrate its position in the array "field(1) field(2) . . ." CSP will not notice if you leave out a field or misspell a name. Subscripts should begin with 1 and go up in ascending order. Please note: The CHANGE command does not work on this panel. For new maps, VARIABLE FIELD EDIT DEFINITION should appear next after pressing PF3.

MAP DEFINITION—VARIABLE FIELD EDIT SCREEN

Once a field has been named, you may assign edit values to the field. The values on this panel (Figure 3.24) are used for two purposes: formatting input and output values and validation of user input. The highlighted (asterisk) field in the bottom portion of the display shows the "current" field. Press ENTER to move to next field (or press PF9, position cursor on desired field, and press ENTER).

Formatting of data includes justification of data, use and position of sign, fill characters, use of currency symbol, display of

```
EZEM26                          MAP DEFINITION

==>
         ENTER = Edit changes (or go to next variable field)  PF3 = Exit
Field Name = fnam1       Map Name = grpnam mapnam      Occurs = 1  Length = 4
..................................... VARIABLE FIELD EDIT DEFINITION ...............................
Data Type        => CHA      Description     => Crayon order number
Justify          => LEF      Decimal Positions  =>      Sign(NO,TRA,LEA)    =>
Fill Character   => N        Zero Edit       => NO  Numeric Separator   => NO
Fold             => MAP      Currency Symbol => NO  Date Edit(1-11)     =>
Hexadecimal      => NO                       Edit Error Message Number:
Input Required   => NO                          Input Required Error   =>
Edit Routine     =>                             Edit Routine Error     =>
Minimum Input    =>                             Minimum Input Error    =>
Minimum Value    =>                             Value Error            =>
Maximum Value    =>                             Data Type Error        =>
Total positions 080                                     Positions   001 to 080
Total lines    024...........................................................Lines    010 to 016
                         CRANDALL'S CRAYON SUPPLY


      CRAYON CODE  ****              COLOR                          #
```

Figure 3.24.   Map definition—Variable field edits.

significant zero, use of commas, conversion of lower-case letters to upper case, and date formatting. Format of input and output data may seem to be desirable, but, it is often best to perform minimal conversion of input data. This allows the application to "see" the data as it was entered. Output formatting is somewhat easier on logic.

Validation of data requires that CSP's programs evaluate user entries. This is either a wonderful feature or a terrible feature, depending upon the nature of your system and the hardware capabilities of your environment. First, CSP's automatic edits can greatly simplify the creation of online applications. Most code written for online programs in COBOL, PL/I, C, or Assembler performs different types of edits. Most of this code is replaced in CSP by using these automatic edits. You even have control over the error messages used for each specific error. Unfortunately, there is a down side. Each time a user responds, CSP will evaluate using the edit criteria specified on this panel. Once CSP finds an error, processing stops and an appropriate error message is returned to the user. In a small system, or in a system where users make few data entry mistakes (most users make few data entry mistakes, by the way), CSP's automatic editing will save huge amounts of effort on the application developer's behalf. But, if a system is large and if the users are prone to error there is a performance penalty associated with the automatic edits. Each time a user responds and is corrected by CSP, a pair of transmissions from the terminal to the mainframe and back must be completed. One of the major causes of poor response time in most large shops is excessive line transmissions. Therefore, it is reasonable to expect that you will be asked to avoid the automatic editing if your users are error-prone and your system is busy. This means that you will either have to code the edits yourself inside CSP processes, or call a non-CSP program to do the edits for you.

Here are explanations of each of the items to be entered on the Variable Field Edit Definition panel.

Data type:

CHA   character data is expected

NUM   only numeric data, sign, and decimal point are allowed

DBCS   only DBCS (double byte character set) data is allowed

MIX   both DBCS and CHA data are allowed

Description:

You may enter up to 30 characters describing the field's contents. Use this documentation opportunity.

Justify:

LEF   Left justify (default for CHA or DBCS data)

RIG   Right justify (default for NUM data)

NO   No justification (CHA, DBCS, and MIX data only)

Decimal positions:

Number of positions allowed (maximum is 18)

Sign:

NO   value displays without sign

LEA   LEAding plus or minus (+ or −) sign may be entered, only negative sign (−) is displayed by system on output

TRA   TRAiling plus or minus (+ or −) sign may be entered, sign is always displayed on output

Field size should allow an extra character for signs when they are used.

Fill character:

This value is used to "pad" incomplete data; key in a value to be used for fill characters. Using the letter N specifies a null value is to be used (CHA and DBCS data default to NULL, NUM data defaults to blank).

Zero edit:

Setting this field to YES forces display of one significant zero for zero amounts.

Numeric separator:

Setting this field to NO or YES specifies whether numeric separators (commas ',') may be entered by the user. This will also cause separators to be automatically displayed by the system (system default separator is the comma, it may be changed).

Fold:

Some terminals allow both lower-case and upper-case letters. This must be supported by the terminal or PC, the controllers, and the network definitions. If the panel displays the value MAP in this field, it means that Fold Input =YES was specified when on the Map Definition Map Specification panel and may not be altered here. Otherwise, setting the field to YES or NO specifies whether text is to be converted to uppercase. Fold=yes indicates that all user entries should be automatically converted to upper-case letters. Fold=no tells CSP to leave data as it is entered by the user.

Currency symbol:

CSP supports the entry of and display of currency symbols via this field. Setting Currency Symbol=YES causes CSP to allow entry of and display the system currency symbol for the field. The currency symbol used is specified by the person who installs CSP.

Date Edit (1-11):

CSP provides several automatic date edit formats. By specifying a number from one to eleven, you may control the format of a date.

1	MM/DD/YY	7	DD/MM/YY
2	MM-DD-YY	8	DD-MM-YY
3	MM:DD:YY	9	DD:MM:YY
4	YY/MM/DD	10	YY-DDD
5	YY-MM-DD	11	YY:DDD
6	YY:MM:DD		

DB2 and SQL/DS users may need to use built-in functions (like CHAR and DATE) to format dates properly. Version 4 of CSP/AD will solve this problem by recognizing DB2 and SQL/DS dates.

Hexadecimal:

Setting this field to YES allows only entry of valid hex digits (0–9, a–f, and A–F) in the field. Blanks and nulls are converted to 0 automatically. Again, try to avoid displaying or accepting entry of hex data if possible.

Input required:

> Sometimes data entry is required in a particular field every time a map is used. Setting Input Required to YES stipulates that a field must be entered. This may be shut off by modifying Input Required to NO.

Edit routine:

> CSP allows the execution of a single code routine as part of the automatic editing it performs. You may either have CSP check the entered data against a CSP table (covered later), or you may have CSP execute a CSP Statement Group (also covered later), or you may have CSP execute one of its internal modulus routines for you. Anyhow, the routine is executed each time CSP evaluates the field.

Minimum Input:

> This field specifies the smallest number of characters that may be entered. For instance, you might require the entry of at least 4 characters out of a possible 30 characters in an address field.

Minimum value:

> Numeric fields may be checked using values input on this panel. This field sets the smallest value allowed.

Maximum value:

> The largest value allowed in this numeric field.

Edit Error Message Number:

> CSP provides stock messages for each edit described previously. You may also use the CSP message facility to create messages for use by the automatic edits. For each of the different types of edits, you may specify the message number to be displayed. You are allowed to specify messages for Input Required Error, Edit Routine Error, Minimum Input Error, Value Error, and Data Type Error.

CSP EDIT SEQUENCE

CSP/AD's automatic field edits take place in the following sequence:

1. Input required
2. Data type
3. Minimum input
4. Minimum value
5. Maximum value
6. Edit routine

Again, each edit takes place every time the user responds.

Once you have finished with the Variable Field Edit Definition panel, press PF3 to exit. You have now finished the basic definition of the map. You may go back and alter the map's definition at any time by choosing the Edit function and selecting the appropriate map function. When editing an already existing map, the panel in Figure 3.25 appears.

The options listed are mostly meaningful, but you should know to choose option 2, Map Presentation, to reenter the "screen painter."

```
EZEM23                          MAP DEFINITION


==>
                ENTER = Edit changes (or go to next field)  PF3 = Exit
                       Map Name = grpnam mapnam
....................................... FIELD ATTRIBUTE DEFINITION ......................................

   NORMAL,DARK,BRIGHT       => NORMAL    NODETECT,DETECT          => NODETECT
   UNPROTECT,PROTECT,ASKIP  => ASKIP     NOHILITE,BLINK,
   ALPHA,NUMERIC            => ALPHA     RVIDEO,USCORE            => NOHILITE
   NOCURSOR,CURSOR          => NOCURSOR  NOMDT,MDT                => NOMDT
   NOENTER,ENTER            => NOENTER   MONO,BLUE,RED,TURQUOISE,
   NOFILL,FILL              => NOFILL    PINK,GREEN,YELLOW,WHITE  => MONO



   Total positions 080                              Positions 001 to 080
   Total lines    024....................................................... Lines 001 to 007
                         CRANDALL'S CRAYON SUPPLY


        CRAYON CODE ****              COLOR
```

Figure 3.25. Map definition—Field attribute definition.

ATTRIBUTES

One vital part of map definition has not yet been examined. Setting attributes on a CSP panel may be done using one of two methods. My favorite is to use the ATTRIBUTE command while looking at the "screen painter" under Map Definition Map Presentation. The other mechanism is to use the Map Definition Field Attribute Definition panel. Both methods will be covered, but first a general discussion of attributes is required.

When a field is defined on a map, a byte is reserved before the field to describe its attribute. The attribute defines the beginning of a field on the screen. Data to be used and/or displayed *follows* this byte. The attribute describes attributes for all bytes that follow until the next attribute byte is encountered. Each attribute occupies one position on the screen. This byte appears as a blank and is not used to store/display data. Attributes come in two basic groups—basic attributes found on all terminals or extended attributes that require specific hardware and software.

Basic 327x Attributes

All 327x-type terminals (includes 327x, 328x, 329x, 317x, 318x, 319x, and others) support certain basic attributes. If you are using a personal computer as a terminal, it uses EMULATION software and/or hardware to mimic one of the 327x terminals. The two basic attributes are intensity (how bright is the data displayed) and protection (may the user overkey the field). The possible values of basic attributes follow:

Intensity:

Normal	Data visible but not highlighted
Bright	Data visible and highlighted
Dark	Data not visible (passwords)

Protection:

Unprotected	Any data may be entered.
Unprotected and Numeric	Any data may be entered. *Most* terminals automatically shift to numeric mode and will allow entry of "non-numeric" data only by pressing the shift key

Protected   Not enterable; if previous field is unprotected, the cursor
                   will pass a protected field only if the user presses the tab
                   key.

Autoskip   Not enterable; cursor will automatically skip over the field

Most screens that contain unprotected fields follow the unprotected field immediately with either an autoskip field or a protected field. The extra field is called a "stopper field," and it limits
the amount of data the user may enter into a specific field to the
desired length. It is also possible for two unprotected fields to
appear together; the attribute byte of the second field ends the
first field. Autoskip is usually used for headings and titles. Protected is usually only specified to delineate numeric fields appearing in a column. This forces the user to enter a TAB key
between each field, reducing the possibility of incorrectly entered
numbers.

Extended Attribute Bytes

IBM provides many terminals capable of "extended" attributes.
In turn, most of the emulators running on personal computers
permit some extended attribute capability. Designing systems
around the use of extended attributes may be risky, since not all
terminals support them and not all software supports them. In
addition, not all people are served well by the use of color (one of
the extended attributes). For instance, many people are
colorblind. Also, an increasing number of terminal users are vi-
sion-impaired and using devices that might not translate well.
Finally, with the proliferation of personal computers being used
as terminals you should be aware that color is not always reliable. Since extended attributes are not supported by all terminals or by all installations, it is sometimes a good idea for
maintenance reasons to avoid them.

Still, the user-friendly capabilities supported by extended attributes often causes organizations to purchase software and
hardware that can support them. A system that uses extended
attributes may be much more pleasant for your users to deal with.
If your installation has invested widely in hardware and software

to support extended attributes, it is probably a good idea to use them if you will have no conflicts with your user community.

Extended Attributes include:

COLOR	Up to eight colors on some terminal types, only four on others. Even if you are using a device capable of more than eight colors, the system cannot support them. Color is an excellent way to highlight entry areas and errors.
UNDERSCORING	Underscoring is useful for showing the user how much data may be (has not been) entered.
REVERSE VIDEO	Is also useful for showing a complete data entry area.
BLINKING	Causes the data in the field to blink. Most users will consider this an *un*friendly act! You may decide to use this feature for certain, very bad errors.
MUSTFILL	When set, the entire field must be keyed before the cursor will advance. No error message is displayed by the system. (This feature is usually considered unfriendly)
MUSTENTER	When used, something must be entered before the cursor will advance. Again, no error message is displayed—the terminal just beeps (also usually considered unfriendly).

If you specify color, then NORMAL and BRIGHT cease to have meaning. The current hardware/software setup does not support colors of varying intensity.

MODIFIED DATA TAG (MDT)

Embedded in the attribute byte is a bit switch called the "Modified Data Tag" or MDT. The MDT is used to tell the terminal that something has been entered by the user (the terminal only sends "changes"). A programmer/developer may trick the terminal into thinking that data was entered in a field by setting that field's attribute to a value that includes MDT "on." When the terminal responds to a poll, only non-null data from fields with MDT "on" (user changed, or program set) is passed to the system. To reduce the amount of data transmitted, new systems should not deliberately set the MDT "on" (either in the definition or by modifying field attributes). If it is required by processing, set the MDT "on" sparingly.

CSP/AD provides two methods to set field attributes, the Map Definition Field Attributes panel (option 6, "Field Attribute Definition" on the Map Definition Function Selection menu), and use of the ATTRIBUTE command while using the screen painter (option 2, "Map Presentation" on the Map Definition Function Selection menu). When you are new to online programming and CSP, the Field Attributes panel is probably the easiest way to go. Once you feel more comfortable with attributes and their settings, you will find it is much faster to use the ATTRIBUTE command while originally defining the map. This will provide you with the added luxury of being able to test the attributes that you set easily.

MAP DEFINITION—FIELD ATTRIBUTES

CSP provides this feature to allow the setting of attributes in a menu-driven manner. New users may find this feature very useful. This panel is reached by choosing option 6, "Field Attribute Definition" on the Map Definition Function Selection menu (see Figure 3.25). It is not part of the sequence of panels displayed while creating a map.

Like the edit definition and naming panels, this panel shows which field is being modified at the bottom of the screen by means of highlighting and asterisk-filling (for uninitialized fields). Press ENTER to advance from one field to the next. Notice that *all* fields are listed, including titles and the spaces following them until the next attribute. This emphasizes the powerful nature of attributes. You may find it useful to define titles with a constant character at both ends. By placing constant markers at both ends of a title, you may then use reverse-video and coloring to their best advantage. If you find moving around with the ENTER key tedious, use the same technique used on the edit and naming panels. This is done using the PF9 (PF21) key to remove/display the CSP/AD portion of this panel (a very useful feature). "Hide" the CSP/AD portion using PF9, place the cursor on the field you want, then redisplay the CSP/AD portion using PF9 again (the chosen field's attributes will be displayed).

The following attribute settings are supported by the Field Attribute Definition panel.

NORMAL, DARK, BRIGHT

Set field intensity

NODETECT, DETECT

Determine if field is sensitive to a light-pen (flashlight-like device used to point to sensitive fields on screen; used mostly in large inventory systems)

UNPROTECT, PROTECT, ASKIP

Sets protection setting

ALPHA, NUMERIC

Sets field to accept character or numeric data (numeric may be overridden by user pressing SHIFT key)

NOCURSOR, CURSOR

Initial cursor position, should be set for first enterable field on the map

NOMDT, MDT

Initial status of MDT for field

NOHILITE, BLINK, RVIDEO, USCORE

Sets one of three mutually exclusive extended attributes: BLINKING, REVERSE VIDEO, and UNDERSCORES (default is NOHILITE, or none)

NOENTER, ENTER

Extended attribute, requires data entry before cursor will move

NOFILL, FILL

Extended attribute, requires that a field be completely filled

MONO, BLUE, RED, TURQUOISE, PINK, GREEN, YELLOW, WHITE

Mono is the default. On color terminals, MONO causes the following colors to be used based upon other attributes:

GREEN—unprotected, normal

RED—unprotected, bright

BLUE—protected/autoskip, normal

WHITE—protected/autoskip, bright

Though extended attributes allow the setting of colors, *be careful!* This may cause a problem due to purchases of new "mono" terminals, terminal emulators that change the colors for you, colorblind users, and other long-range problems.

NOUTLINE, OLEFT, ORIGHT, OOVER, OUNDER, BOX

This setting allows use of the IBM 5550 terminal's capability to draw lines around data fields on the screen. The default is NOUTLINE since most terminals do not support this feature. It is not unusual for those installing CSP to suppress display of this option entirely.

An attractive alternative to using the Field Attribute Definition panel is to return to the map presentation display and use the ATTRIBUTE command.

SETTING ATTRIBUTE WITH A COMMAND

The ATTRIBUTE command may only be used from the "screen painter." The screen painter is reached by choosing option 2, Map Presentation, from the Map Definition Function Selection panel. The settings for the ATTRIBUTE command are the same as those on the Field Attribute Definition panel, though they must be spelled out in the command. The syntax for the ATTRIBUTE command is shown in Figure 3.26. Basic attribute options include: ALPHa or NUMeric, BOX, CURsor, DARK or BRight or NORMal, DETect or NODETect, Protect or UNProtect or ASKIP, MDT or NOMDT.

Extended attribute options include ENTer or NOENTer, FILl or NOFILl, MDT or NOMDT, MONOchrome or BLUE or GREen or PINK or RED or TURQuoise or WHite or YELlow, NOUTline,

```
ATTribute      ADD or +
               ALPHa or NUMeric
               BOX
               CURsor
               DARK or BRight or NORMal
               DETect or NODETect
               MDT or NOMDT
               Protect or UNProtect or ASKIP

               ENTer or NOENTer
               FILI OR NOFILI
               MONOchrome or BLUE or GReen or PINK
                     or RED or TURQuoise or WHite
                     or YELlow
               NOHIlite or NOHighlight or REVersevideo
                     or RVIDEO or UNDerscore or UScore
                     or BLinking
               NOUTline or OLEFT or ORIGHT or OOVER
                     or OUNDER
```

Figure 3.26.   ATTRIBUTE command syntax.

OLeft, ORight, OUnder, OOver, REVersevideo or RVIDEO or UNDerscore or UScore or BLInking or NOHIghlight or NOHIlite.

Again, the command works only on the map presentation display (screen painter). Use &ATTR when repeating the command for several fields. Keying in the command ATTRIBUTE alone and pressing enter will display the attribute values for the field where the cursor currently resides.

To see the current attribute setting for a field, do the following:

1. Key in: ATTRIBUTE.
2. Move cursor to field (use keyboard TAB key).
3. Press ENTER.

To completely change attributes for a field, enter the ATTRIBUTE command followed by the desired values, place the cursor on the desired field, and press ENTER.

To set the attributes of a field to two values, do the following:

1. Key in: ATTRIBUTE value1 value2.
2. Move cursor to field.
3. Press ENTER.

To modify or add attributes, the plus sign (+) is used. Enter the ATTRIBUTE command followed by a plus sign and the desired new values, place the cursor on the desired field, and press ENTER.

Do the following to add attributes to a field:

1. Key in ATTRIBUTE + value1 value2.
2. Move cursor to field.
3. Press ENTER.

CHAPTER REVIEW

At this point, you have been introduced to the features of map definition. Here is a list of suggestions.

- Design the map first.
- Set the correct device types for later display.

- Define the map using Constants and Variables.
- Use Spacers to ease panel definition.
- Consider using the ATTRIBUTE command to set attribute values.
- Name all fields uniquely; name the message field EZEMSG.
- Use the TEST FIL command to view your map.
- Don't forget to use PF9/PF21 to "hide" CSP information occasionally.

CHAPTER 3 EXERCISES

Questions

1. What is the term used to describe a CSP output screen?
2. What is the term used to describe several CSP output screens defined together for use by an application?
3. What character is used to begin a constant field in your system?
4. What character is used to begin a variable field in your system?
5. What is the easiest way to center data on the screen?
6. Which command is used to view the screen as it is defined?
7. Which key is used to CONTINUE creation?
8. Which fields must be named?
9. What is the special name that should be used for messages?
10. Name both ways that attributes may be set.
11. How many fields should use the attribute CURSOR? Which one(s)?
12. Why should you be cautious about the use of colors?

Map Definition

1. Design a screen that will display the following fields:

Inventory ID	12	Characters (unprotected), required
Title	40	Characters (auto skip)
Author	40	Characters (auto skip)
Publisher	20	Characters (auto skip)

ISBN Number	13	Characters (auto skip)
Replacement cost	8	Positions (unprotected, numeric) (Between $0.00 and $500.00 inclusive)
Sale price	8	Positions (unprotected, numeric) (Between $0.00 and $999.99 inclusive)
Quantity on hand	6	Characters (auto skip)
Quantity on order	6	Positions (unprotected, numeric) (Between 1 and 100 inclusive)

Your screen should also have a title at the top (be creative) and an error message field at the bottom. Be sure to use EZEMSG as the message field name.

2. The map group name is BB01G; the map name is BB01M01.
3. Make all fields normal intensity *except* the title and the message. If you are working on a color-capable system, try using different colors for different items.
4. Use the CSP/AD TEST command to see what your map will look like. Don't forget, TEST will not activate any built-in edits you may have defined.

4

Basic Application
Definition and Testing

CSP "programs" exist as something called APPLICATIONs. Each application is made up of several routines called PROCESSes. Application processes can do everything a normal program can do, usually with far less code. CSP processing is sometimes knocked as being less efficient than older, more traditional code. CSP processing is not inherently inefficient, but the ease with which an application can be generated lends itself to sloppiness by the application designer. The ease of creation, coupled with CSP's internal processing, can create a truly impressive consumer of system resources. Proper system design and close attention to the runtime environment will allow *most* (maybe 80 percent) of a given system's applications to run effectively using CSP. Use of CSP can result in significant savings overall due to greatly reduced application development and maintenance time.

Like map definition, CSP applications are created using a series of panels. Each panel provides a different piece of the application's definition. As in the map definition, you move from one panel to the next. Be careful! Sometimes you press ENTER and other times you press PF3 to continue. For existing applications, you must select which function you would like to modify. At the end of this chapter you will be asked to design and define a CSP application consisting of several processes.

SCREENS INVOLVED IN APPLICATION DEFINITION

CSP/AD provides several screens as part of the application definition process. Each of the different screens is listed below with an explanation of their function.

Definition Selection:

Used when modifying existing CSP objects. This panel shows a menu containing choices leading to the major CSP object types.

Application Definition Function Selection:

After selecting option 5 from the Definition Selection panel, this panel provides another menu. Each option on the Function Selection menu corresponds to different parts of the application to be modified. This panel does not appear during the initial creation of an application.

Application Specifications:

This panel describes the type of application and names several optional objects to be connected to the application. The types of optional objects that may be named include main working storage record, map group, message file, help map group, and DL/I PSB. Function keys may also be specified for help and edit bypassing. The panel also allows specification of whether 12 or 24 PF keys are to be supported and implicit definition of variables.

Process list and definition:

Main processes are defined on this panel. In most modern systems only one main process exists, a "mainline" process. Pressing PF4 from this panel causes an expansion of the application's structure by listing all lower-level processes. From this panel the code of individual processes and objects may be selected for editing.

Table and Additional Records List:

If an application will use CSP tables or records other than the main working storage record defined as part of the Application Specifications, this panel is used. Each of the tables or records is listed by name along with a code indicating whether it is a table or record.

Called Parameter List:

Used by Called Transaction/Batch applications to list the parameters passed into this application from outside. Each item listed must be represented by a corresponding data item or record defined to the application.

Structure List:

This is very similar to the Process List except that Main processes may not be defined. Another difference is that lower-level processes are automatically displayed on this panel. Pressing PF4 from this panel refreshes the display of lower-level processes. From this panel the code of individual processes and objects may be selected for editing.

Application prologue:

An opportunity to document the application. This information will be printed every time the application is. Don't miss this built-in opportunity to record your intentions!

APPLICATION DESIGN

The ease of creating applications in CSP often causes developers to simply "throw something up" to see if it works. Unfortunately, if it partially works, no matter how poorly, we don't seem to have time to fix it. Take the time to design your CSP applications! Online design can be as simple as the old INPUT-PROCESS-OUTPUT model you learned in DP-101.

INPUT

Describe input files/databases

Describe desired user inputs/responses/outputs.

PROCESS

What is to be done with the files/databases?

How will user inputs/responses be handled?

OUTPUT

What is the output from the process?

What do the outputs look like?

Where do the outputs go?

It is best to follow your installation's design methodology or use the available CASE tools to design your application. In any case, this is not a step you can afford to miss.

Use the PROLOGUE to document your intentions. Remember, *you* may be asked to fix a bug in this system years from now. Make sure that the design is modular and detailed at least to the point where individual I/O activities are identified. Each I/O activity will require execution of a CSP PROCESS OPTION. Other processes will use either a CSP EXECUTE PROCESS OPTION or something called a STATEMENT GROUP. Create paper documentation that shows the following:

- samples of all inputs
- samples of all outputs
- explanation of processing and logic
- relationships between subprocesses (structure chart in Figure 4.1)

DEFINING THE APPLICATION TO CSP

The steps to define an application to CSP are simple.

1. Design the application.
2. Define the application to CSP using the available screens (be careful not to add unnecessary processes). The mainline logic and all I/O routines must be defined using CSP PROCESS OPTIONs. All other logic will either be included inside the

logic of the mainline EXECUTE PROCESS OPTION or as part of the I/O PROCESS OPTIONs and their logic in separate EXECUTE PROCESS OPTIONs or STATEMENT GROUPS.

3. Decide which process options to use to match your design. Each PROCESS OPTION execution follows the following logic path:

- statements executed BEFORE the process option is executed
- process option execution
- statements executed AFTER the process option is executed
- flow to next process
  (Flow represents fall-through or GOTO logic between the application's MAIN processes. It is not used in most shops due to conflicts with "structured programming" philosophy.)

The rest of this chapter shows the specific screens used to create and test a CSP application. If you have created a "structure chart" or used another tool to illustrate your modularized design, each program function will be represented by at least one PROCESS.

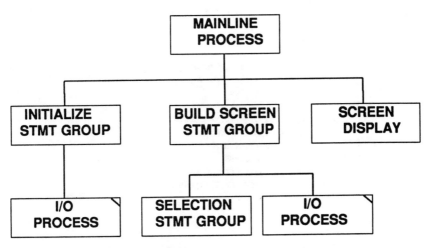

Figure 4.1. Structure chart.

DEFINITION SELECTION

Once again, when you begin the editing process, CSP/AD provides you with a choice of the type of object to edit. For new applications, enter the new application's name and enter a 5 as the member type before pressing ENTER. When you create a new object, the system will "walk" you through a series of panels. Use the PF3 key to complete one stage of the definition and proceed. If your object already exists, you simply name it and press ENTER. Since objects must be uniquely defined, CSP will find the application definition and present a Function Selection panel next. (The Function Selection panel will not appear when creating a new application.) See Figure 4.2.

Application names must follow these specific rules:

- no longer than seven (7) characters long
- must begin with a letter (A–Z)
- remaining characters must be alphanumeric (A–Z, 0–9)

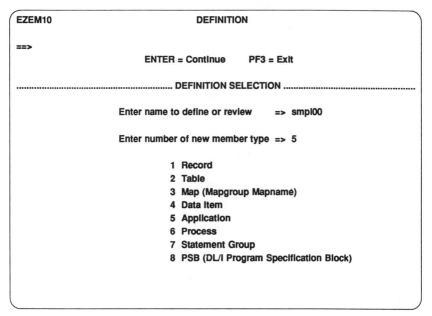

Figure 4.2.   Definition selection.

- no embedded blanks
- first three (3) characters may not be "EZE"

Press ENTER to advance to the next screen.

APPLICATION DEFINITION—
FUNCTION SELECTION

This panel does not display during the initial creation of an application. The Function Selection panel will display when you wish to review or edit an existing application. Key in the number corresponding to the function desired and press ENTER (Figure 4.3).

- Process List and Definition—names the processes that are part of the application and their relationship to one another (allows addition/deletion of main modules, may display structure with PF4)
- Table and Additional Records List—lists tables used for editing and supplemental working storage records
- Called Parameter List—names parameters to be used by called processes

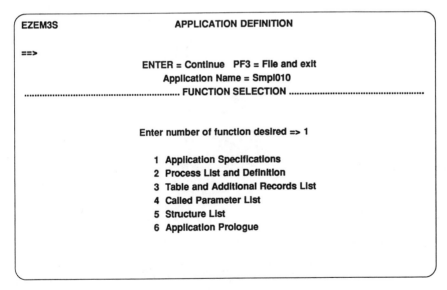

Figure 4.3.   Application definition—Function selection.

- Structure List—illustrates relationships of various subprocesses that are part of this application, should match design (useful when viewing completed application but does not allow addition/deletion of main modules)
- Application Prologue—used to document application processing

Press ENTER to continue with the application definition.

APPLICATION SPECIFICATIONS

Each application is defined to be of a particular type and is associated with specific working storage records and maps. The Application Definition Application Specifications panel (Figure 4.4) is where these important relationships are defined. In addition, the name of the message file, special function keys, implicit definition of data, and IMS initial screens are defined here.

```
EZEM31                         APPLICATION DEFINITION

==>
                    PF3 = Exit (or continue if new definition)
                    Application Name = Smpl010
.............................................. APPLICATION SPECIFICATIONS ...........................................


    Type of application => 1
                                     Working Storage  => WMPL00

      1   Main Transaction
      2   Main Batch           Map Group Name => grpnam
      3   Called Transaction
      4   Called Batch         Message File    =>

    Help Map Group Name  =>    PSB Name        =>

    Help PF Key        =>      PF1-12=PF13-24  => YES

    Bypass Edit PF Keys  =>    Allow Implicits  => YES

                               First Map        =>
```

Figure 4.4. Application definition—Application specifications.

The following items may be entered on this panel.

Type of application

Main Transaction (only choice that may use maps)

Main Batch (no maps, not frequently used)

Called Transaction (no maps allowed)

Called Batch (no maps, not frequently used)

CSP is not frequently used for batch applications. CSP Version 3 Release 3 offered many performance enhancements, most notably the ability to generate and execute VS COBOL II code. Version 4 will make this option even more attractive.

Working Storage

Names main Working Storage record used

Map Group Name

Names Map Group used by application (all maps used must be part of this group)

Message File

Three character message file ID (DCA___D), plus optional one character application ID (more on this later)

Help Map Group Name

Names Map Group containing HELP maps

PSB Name

Identifies DL/I PSB to be accessed

Help PF Key

Names HELP key for this application (default=PF1)

PF1-12=PF13-24

Makes PF13-24 the same as PF1-12, this is useful for terminals where PF1-12 require use of the ALT key but the keypad holds PF13-24

Bypass Edit Keys

> Names keys used to bypass internal edits, necessary if using a key for "exit" capability (e.g., " => 3,12" makes PF3 & PF12 bypass keys)

Allow Implicits

> CSP creates variables used but not defined (not usually a good idea). This does not work consistently across environments.

First Map

> IMS only, identifies map upon which input is performed before executing the first process

All fields on this panel are optional except application type. Main Transactions require a Map Group. Most applications use a working storage record and it should be named here. If CSP's automatic messaging will be used (see Chapter 9), you must identify the message file. Bypass keys are necessary if the application will provide users with the ability to exit an application or transfer to another application using function keys. If the chosen exit or transfer keys are not listed as bypass keys, CSP will pass control to the application only when all edits are successfully passed. Press PF3 to continue the definition process from here.

CSP PROCESS OPTIONS

CSP applications (programs) are based upon the execution of one or more functional PROCESSes (modules). Each CSP process follows a standard execution path, allowing the application developer to insert code both BEFORE and AFTER the main function of the process. Most CSP processes are based around I/O functions, either map I/O or record I/O. Process functions are described by naming a PROCESS OPTION at the time the process is defined. For example:

- A process called "sendmap" might invoke the CONVERSE PROCESS OPTION to send and receive a map.

- A process called "getrec" might invoke the INQUIRY PRO-CESS OPTION to retrieve data from a database.
- EXECUTE PROCESS OPTION processes perform no I/O.

Processes are executed in the following two basic fashions:

- PERFORMing them from within other processes (the approved method)
- Falling-through or FLOWing to the process (not generally a good idea)

File and database PROCESS OPTIONS also require that the application developer designate an ERROR routine. If an error routine is not specified for file/database-oriented processes, CSP will abend the application when a non-zero condition results from the I/O.

BASIC FORMAT OF PROCESS DEFINITION

The Application Process List (Figure 4.5) is used to define MAIN processes for the application. Most applications require only one main process, analogous to a "mainline" routine when using COBOL or PL/I. All other processes or statement groups are performed through logic in the main process or by one of its subordinates. When returning to this panel after adding CSP

```
EZEM36                      APPLICATION DEFINITION

==>
              ENTER = File and continue        PF3 = File and exit
              PF4 = Display application structure
                     Application Name = SMPL010
Select Definition:   S = P+F+L   P = Processing  F = Flow   E = Edit Object
              O = Object Selection            L = Structure List
Total lines 0001 ........................APPLICATION PROCESS LIST .......................................
SEL PROCESS                   OPTION          OBJECT              ERROR
***                           TOP OF LIST
001 smpl010                   CONVERSE        mapnam              errgrp
***                           END OF LIST
```

Figure 4.5   Application definition—Application process list.

statements to perform processes or statement groups, press PF4, and the new process names (used in the perform statements) will appear on the Application Process List.

The information listed on this panel includes the following:

PROCESS

Unique name given to a process

1–18 characters (A–Z, 0–9, $, #, @, –, _)

(limited to 7 characters without – or _ before CSP V3 R3)

first character must be (A–Z, $, #, @)

no embedded blanks

first three characters may not be EZE

OPTION

CSP PROCESS OPTION to be performed (see following pages)

OBJECT

Name of record or map PROCESS OPTION I/O will use

(not used for EXECUTE or SQLEXEC process options)

ERROR

Name of the error routine to be executed if errors occur while performing PROCESS OPTION. Error routine should be a statement group or one of the special CSP names EZECLOS, EZEFLO, or EZERTN (explained later). If you have decided to use the FLOW (GO TO) capabilities of CSP, you may also name a "main" process if invoked as part of a FLOW (no-no!).

Omitting error routines causes the application to end and issue an appropriate error message should an error occur.

FLOW

Names the process to be executed next. If no process is named, control passes to the next process in this application.

Flow implies "GO TO" processing, something that is avoided in most modern programming designs. It is probably best to avoid or strictly control use of flow processing.

The leftmost column of the Application Process List is used to control which part(s) of a PROCESS you wish to define or edit. "S" means that you will enter Processing logic first, then enter Flow logic, and finally the system refreshes the Structure List upon exit to show any processes or statement groups you performed (this is the default for new processes). "P" indicates that you merely wish to enter Processing logic. This is the choice most frequently used when creating/editing processes. Be sure to press PF4 to refresh the Structure List after entering process logic. Pick a name for your mainline process, use a process option of EXECUTE, and press ENTER. "F" allows you to modify only the Flow portion of the process (rarely used). Later, once record or map processing processes have been defined, "E" may be used to edit the record or map definition. For SQL processes, "O" allows access to the SQL code used by the process. Selection choice "L" expands the structure beneath a specific module; it is easier to press PF4 and refresh the entire Structure List. Make sure a desired code is in front of the procedure you wish to edit, and press ENTER.

STATEMENT DEFINITION SCREEN

After using a "P" or "S" from the Application Process List for a process, the Application Process Statement Definition panel appears (Figure 4.6). If an I/O process is involved (CONVERSE, INQUIRY, and so on), the part of the panel where statements are keyed will be divided by a PROCESS OPTION line. Code entered above this line will be executed BEFORE the I/O; code entered below this line will be executed AFTER the I/O and any ERROR routines have executed. If the routine being edited is an EXECUTE process or a STATEMENT GROUP (defined later), then all code occurs at once.

Code the CSP statements for the process being defined in the space provided. CSP/AD "line commands" may be used to insert, delete, copy, move, or repeat rows (see Appendix A for syntax).

```
EZEM39                    APPLICATION PROCESS DEFINITION

==>
                  PF3 = File and exit (or file and continue if more selected)
        Process => SMPL010    Description = Test EZEAID for BYPASS
        Option = CONVERSE           Object = mapnam
Total lines 0009 ...............................STATEMENT DEFINITION ...........................................
***                             TOP OF LIST
001  SET MAPNAM CLEAR ; note that comments may follow semi-colon
*** ------------------------------- PROCESS OPTION -------------------------------
002  IF EZEAID IS BYPASS      ;   this is a test for the bypass key
003       IF EZEAID IS PF3    ;   if bypassing because of PF3
004            EZERTN         ;       exit & return
005       END                 ;       ends inside if
006       IF EZEAID IS PF12   ;       if bypassing because of PF12
007            EZECLOS        ;       exit application
008       END                 ;       ends 2nd inside if
009  END                      ;   ends 1st if
***                             END OF LIST
```

Figure 4.6.   Application process statement definition.

Many of the available statements are discussed in Chapter 5. Key in the statements to be executed BEFORE the process option above the line marked PROCESS OPTION, which illustrates where the I/O will actually take place in the logic (not present for EXECUTE processes or STATEMENT GROUPS). Key in the statements to be executed AFTER the process below the line marked PROCESS OPTION. Use semicolons (;) to delimit statements. (This was optional and inserted automatically prior to CSP V3 R3, but it was required by CSP V3 R3 and later versions.) Statements may extend over multiple lines. Text following a semicolon is treated as a comment. An entire line may consist of comments if the first character is a semicolon. The next several pages provide an example of coding a mainline process and a subordinate process for displaying a map.

DEFINING A "MAINLINE" PROCESS

In most applications, only one module is defined at the highest level. This is a so-called "mainline" routine that causes every-thing else to happen. See Figure 4.7.

```
EZEM36                    APPLICATION DEFINITION

==>
              ENTER = File and continue        PF3 = File and exit
              PF4 = Display application structure
                   Application Name = SMPL010
Select Definition:  S = P+F+L  P = Processing  F = Flow  E = Edit Object
                   O = Object Selection        L = Structure List
Total lines 0001 ........................APPLICATION PROCESS LIST .....................................................
SEL PROCESS                    OPTION          OBJECT              ERROR
***                            TOP OF LIST
001 mainlin                    execute
***                            END OF LIST
```

Figure 4.7.   Application definition main process.

On the first line, key in a process name. *Be careful!* Processes become members in the MSL; they use unique names for the "mainline" routines of different applications unless they are meant to share the "mainline" code (not a normal circumstance). Process names may be up to 18 characters long and must follow these rules:

1. The first character must be alpha or national (A–Z, $, #, @).
2. Other characters may include numbers, hyphens, and under-scores (A–Z, 0–9, $, #, @, –, _).
   Note: Generation of VS COBOL II code causes characters not valid for COBOL names to be replaced by hyphens.
3. Names may not contain embedded blanks.
4. Names may not begin with EZE.

"Mainline" processes should use the EXECUTE process option. Since no I/O is performed by the EXECUTE process option, neither object nor error routines are specified. Notice that the CSP automatically chooses the "S" definition option. Replace it with a "P" and press ENTER (or just press ENTER and ignore the FLOW panel later). When the Application Process Statement Definition panel appears (Figure 4.8), key in the application's control statements. Code the CSP statement to perform a CON-VERSE routine name "sendmap" as illustrated in Figure 4.8. You will find that names like this (and "mainline"), while de-

```
EZEM39                    APPLICATION PROCESS DEFINITION

==>
                  PF3 = File and exit  (or file and continue if more selected)
        Process => SMPL010    Description =
        Option = EXECUTE         Object =
Total lines 0009 ...............................STATEMENT DEFINITION ............................................
  ***                         TOP OF LIST
001   perform sendmap ; note that comments may follow semi-colon
002   EZECLOS          ;exit from application
  ***                         END OF LIST
```

Figure 4.8.　　Application process statement definition.

scriptive, cause problems in a shared MSL. It is best to use a naming convention like that first introduced in Chapter 2. If an application name is "XX01A," then processes are often named "XX01Pxxxx" so that they share a common name. This isn't pretty, but it will allow the greatest flexibility when creating new applications or maintaining existing ones. Statement syntax is covered more completely beginning in Chapter 5. End the application using the EZECLOS statement (you should know, however, that CSP automatically places an EZECLOS at the end of every application execution, so this is redundant). After entering the statements, press PF3 to save and return to the PROCESS LIST.

Upon returning to the Process and Group List panel, press PF4 to display the APPLICATION STRUCTURE (this is automatic if the Selection Definition option chosen earlier was "S"). This panel (Figure 4.9) also appears after using option "L." When CSP detects the PERFORM statement in the mainline logic, it automatically adds the performed module to that application structure (when displayed with PF4). Note the relative level numbers of the modules (LVL column). Level one is the highest level in an application's logic. Lower levels illustrate subordinate logic. When you are finished, the module names and level numbers displayed after pressing PF4 here should match your application design.

Alter the OPTION for the SENDMAP to CONVERSE and key in the name of the map to be displayed under OBJECT (CSP

```
EZEM37                         STRUCTURE LIST                           More->

==>
            ENTER = File and continue        PF3 = File and exit    PF4 = Refresh
                               Member name SMPL010
Select Definition:  S = P+F      P = Processing    F = Flow E = Edit Object
                    O = Object Selection           Maximum level => 2
Total lines 0001 ......................PROCESS AND GROUP LIST ......................................
SEL NAME                    LVL  OPTION        OBJECT    ERROR
***                              TOP OF LIST
001 MAINLIN                 001  EXECUTE
P-> SENDMAP                 002  CONVERSE   mapnam
***                              END OF LIST
```

Figure 4.9.   Structure list.

defaults option to EXECUTE). CSP obtains the map group name from the application definition panel. CONVERSE causes the named map to be displayed after processing logic coded BEFORE the process option, allows the user to respond, then does the logic coded AFTER the process option. CONVERSE does not require an ERROR routine. Since this application does not call for any map handling logic (we're simply going to display the panel), no code is necessary inside the CONVERSE routine.

To add descriptions, press PF11 (to shift right, note "More" arrow at top of page) to see the description column. Press PF10 to return to this panel. Press PF3 to save the application's process definitions. The application is now ready to test.

CSP/AD TEST FACILITY

CSP/AD provides an online testing facility for debugging applications. This facility is not available in the current Programmable Workstation product, but it will be in the fall of 1992. CSP's testing tool incorporates a trace that shows each statement executed, each branch, and the contents of variables when tested or changed. Customizing the test environment is easy. You may choose to trace only selected processes or view selected data. Break points may be established at individual lines of code or routines. Data viewing and modification at breakpoints helps ease the debugging process.

The test facility is entered via option 3 from CSP/AD's "main menu," the Facility Selection menu. You may also enter test by using the "=Test" facility transfer command (see Chapter 7) or from the LIST facility (also covered in Chapter 7).

Usually, applications are nearly identical when testing and under CSP/AE with two significant exceptions. First, CICS's pseudoconversational (CSP segmentation) is not fully simulated. Second, if you choose to ALLOW IMPLICITS, the test facility will implicitly create undefined variables. When generating an application, any undefined variables will show up as an error.

Using the Test Facility

Again, the test facility is entered via option 3 from CSP/AD's "main menu," the Facility Selection menu. You may also enter test by using the "=Test" facility transfer command or from the LIST facility (both covered in Chapter 7). Setting up the test environment is a process begun on the very first test facility screen. The tool is very flexible but requires that you specifically set the options on this panel each time you begin testing.

The data to be entered on this panel (Figure 4.10) is listed below:

1. Enter name of application to be tested (must be in current MSL concatenation).
2. Choose testing function. Most of the time, option 2, RUN, is desired. Large applications undergoing initial testing should probably be run through the syntax check at least once to identify basic errors.
3. Specify run options desired on this panel and on the next panel displayed. Some of these choices will cause other panels to be displayed before beginning the testing process. It is important in any case to specify a Stop Key.

Trace => YES or NO

Usually desirable, turns on the trace; you may wish to shut this off or limit the trace when testing (see Figure 4.11).

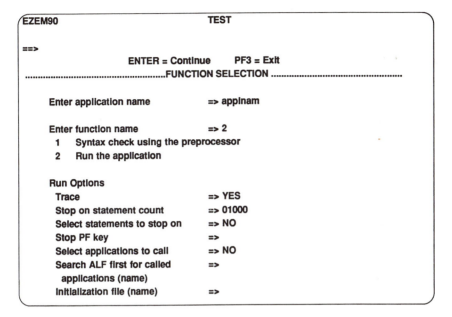

EZEM90   TEST

==>
                    ENTER = Continue   PF3 = Exit
....FUNCTION SELECTION ....

    Enter application name   => applnam

    Enter function name   => 2
      1   Syntax check using the preprocessor
      2   Run the application

    Run Options
      Trace   => YES
      Stop on statement count   => 01000
      Select statements to stop on => NO
      Stop PF key   =>
      Select applications to call   => NO
      Search ALF first for called   =>
         applications (name)
      Initialization file (name)   =>

Figure 4.10.   Test—Function selection.

Stop on statement count => n

Causes automatic stop after n statements

Select statements to stop on => YES or NO

Allows setting breakpoints (see Figures 4.18 and 4.19)

Stop PF key => x

Enter the PF key's number (e.g., 9 for PF9) that will be used to halt testing if the module is looping. It is important to use this feature EVERY TIME you test. This is the most reliable way to stop a test.

Select applications to call => YES or NO

Causes only selected applications to be CALLed; this can cause problems if the CALLed routine's processing is integral to the test.

Search ALF first for called applications (name) => alfdd

Causes the named ALF to be searched for CALLed routines rather than interpreting the routines from the MSL

Initialization file (name) => ddname

File used to store initialized working storage record, useful when testing routines that will be CALLed, XFERed to, or DXFRed to

After responding to all choices on the Test Function Selection panel, press ENTER, and the Trace Options panel will display.

The panel in Figure 4.11 allows you to control the level of tracing. These values may be reset at stop points using a similar display.

Trace to terminal => YES or NO

If sending trace data to a file, it may be quicker to test by shutting off the terminal display.

```
EZEM91                          TEST

==>
                ENTER = Continue              PF3 = Exit

.......................................... TRACE OPTIONS ........................................


             Trace to terminal         => YES
             Trace to file (name)       =>
                Fold to uppercase       =>
             Select processes to trace  => NO
             Select data items to trace => NO
             Select trace data          => NO




```

Figure 4.11.   Test—Trace options.

Trace to file (name) => alfdd.member

Causes trace data to be sent to the named ALF and member. If the ALF's ddname is omitted, the member is added to FZERSAM.

Fold to uppercase => NO or YES

Causes file-directed trace output to be converted to uppercase

Select processes to trace => NO or YES

Allows trace of only selected processes (see Figure 4.15)

Select data items to trace => NO or YES

Allows selective tracing of data items (see Figure 4.16)

Select trace data => NO or YES

If yes, the trace may be customized (see Figure 4.17).

If none of the Select options are turned on, testing begins. For each Select option chosen, a different panel is displayed to allow further specification (Figures 4.15 through 4.19).

When testing begins, CSP checks to see if any file(s) are to be used and whether they have been allocated. If files are to be used, a panel similar to the one in Figure 4.12 appears. If the name associated with a record's FILE NAME (see record definition in Chapter 6) is already allocated to CSP's test environment, the file name and system name will be displayed for your review. If the name associated with the record's file name is not already

```
EZEM95                        TEST
EZE00350A Enter system resource names or press ENTER for defaults
==>
                    ENTER = Continue  PF3 = Cancel
Total lines 001 ............................. TEST FILE LIST ......................................Lines 0001 to 0001
FILE NAME              SYSTEM RESOURCE NAME           CMS FILE  RECORD
smplrc                test.flle                        NO       smplrc
```

Figure 4.12.   Test—Test file list.

allocated, you may enter the system name for the test file to be used. The four fields displayed may only be entered if CSP does not find the file preallocated.

FILE NAME

Defined as part of record definition, this name is a DDNAME in MVS and a DLBL under VM.

SYSTEM RESOURCE NAME

Actual name of file on system, DSN under MVS (no quotes) or VM File Name and File Type.

CMS FILE

Must be marked YES when file is CMS.

RECORD

Name of CSP record definition being used

In order for CSP to find the files preallocated, they must be described to the system BEFORE entering CSP/AD. Many systems provide a panel before entering CSP/AD that allows definition of various files. If not, the commands to be used are system dependent. Figure 4.13 illustrates the TSO and VM/CMS commands necessary to allocate a file. These commands would be entered BEFORE entering CSP/AD. If you are unfamiliar with the appropriate command for your environment, speak with your

TSO

```
ALLOC dd(smplrc) da('test.file') shr reu
```

VM/CMS

```
DLBL smplrc a CMS test file
```

Figure 4.13. TSO and VM file allocation statements.

CSP administrator. The next several pages detail different testing panels you might encounter.

Test—Run Options

The RUN OPTIONS page is displayed at stop points and may be redisplayed at any testing stop point by pressing PF4 while stopped (Figure 4.14). The options on this panel are the same as those entered at the beginning of the testing process (Figures 4.10 and 4.11). In this way, you may dynamically alter the test output based upon which stop point you have encountered.

```
EZEM99                              TEST

==>
                  PF3 = Continue              PF4 = Return to Stop Screen
                  Application = applnam       Process or Group = xxxxxx
        ...................................... RUN OPTIONS ................................................

            Trace Options
                Trace to terminal            => YES
                Trace to file (name)         =>
                 Fold to uppercase           =>
                Select processes to trace    => NO
                Select data items to trace   => NO
                Select trace data            => NO

            Stop Options
                Stop on statement count      => 01000
                Activate statement stops     => NO
                Select statements to stop on => NO
                Stop PF key                  => 10
```

Figure 4.14.   Test—Run options.

Test—Select Processes to Trace

Specifying "Select processes to trace = yes" on the either the TEST RUN OPTIONS or TEST TRACE OPTIONS panels causes the screen in Figure 4.15 to display. Key an "S" next to the process(es)/group(s) you wish to trace; no tracing will be done for

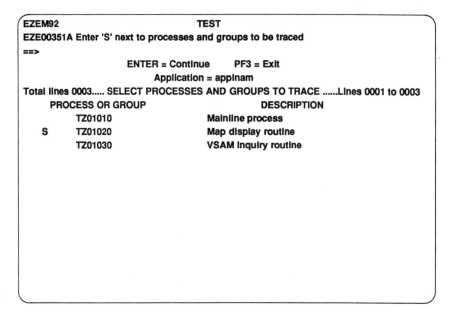

Figure 4.15. Test—Select processes and groups to trace.

unmarked processes/groups. You must press ENTER after selecting processes and groups to trace. Pressing PF3 immediately
(no processes/groups marked with "S") will cause all routines to
be traced (the default).

Test—Select Data Items to Trace

Setting "Select data items to trace = yes" causes the screen in
Figure 4.16 to display. Choose items to be traced by placing an
"S" in front of them. The selected items will be displayed each
time their value changes. As part of CSP/AD application definition, several different objects are used that reference data items.
For each object that contains data items (record, map, and so
forth) this panel is displayed. The first list of items displayed is
the set of EZE variables (EZE connotes a special CSP word).
Each time you press ENTER, CSP will display variables from a
different object (map or record) for your selection. Selecting no
data items causes CSP to default to tracing each data item that
changes while testing. Again, press PF3 to exit.

```
EZEM93                              TEST
EZE00352A Enter selection "S" of Items for data tracing
==>
                    ENTER = Continue      PF3 = Exit
              Application = applnam    Object =
Total lines 0004 ............... SELECT DATA ITEMS TO TRACE .................Lines 0001 to 0004
      EZETST        EZEMNO        EZEDTE        EZEDAY        EAETIM
      EZEUSR        EZEAID        EZEERM        EZERT2        EZERT8
  s   EZEAPP        EZESEGM       EZEOVER       EZEOVERS      EZEFEC
      EZECNVCM      EZESEGTR      EZERCODE
```

Figure 4.16.   Test—Select data items to trace.

Test—Select Trace Data

Specifying "Select trace data = yes" causes the display in Figure 4.17 to appear. Choose the data you wish to see while tracing. See Figure 4.20 for a sample of trace output.

```
EZEM9B                              TEST

==>
                    ENTER or PF3 = Continue
                    Application = applnam
...............................................SELECT TRACE DATA.................................................

          Warning messages          => YES
          Path labels               => YES
          File I/O                  => YES
          Statements                => YES
          Data content              => YES
          Call data                 => YES
```

Figure 4.17.   Test—Select trace data.

Warning messages =>> YES or NO

Controls display of CSP warning messages

Path labels =>> YES or NO

Displays GROUP, PROCESS, CALLed APPLICATION, and FLOW names along with direction arrows (from —> to) each time a branch occurs

File I/O =>> YES or NO

Shows results of I/O by listing contents of buffers after the I/O occurs

Statements =>> YES or NO

Lists CSP statements as they are executed (statements are marked in the trace output by following a series of "greater than" symbols ">>>>")

Data content =>> YES or NO

Shows contents of data items after data changes

Call data =>> YES or NO

Lists contents of items being passed by a CALL

STOP OPTIONS—CONTROLLING THE TEST

Once you have begun testing an application, it will execute until finished or an error has occurred, or halted by the user in some way. TESTing may be halted by the user in one of these four ways:

1. Press the PF key defined as STOP key on first TEST screen. This may be done whenever the "Press Enter to Continue" message is displayed.
2. Press PA2 when the "Press Enter to Continue" message is displayed.

```
EZEM98                        TEST
EZE00660A Specify the desired action next to the process or group

==>
                   PF3 = Continue        PF4 = Next Application
S = Stop before the first statement in the process or group
L = List statements for selection
C = Cancel statement stops in the process or group
                      Application = appinam
Total lines 0003 ..........SELECT PROCESSES OR GROUPS TO STOP IN........... Lines 0001 to 0003
   PROCESS OR GROUP        STOPS SET              DESCRIPTION
       TZ01010               NO              Mainline process
  S    TZ01020               NO              Map display routine
       TZ01030               YES             VSAM inquiry routine
```

Figure 4.18. Test—Select processes or groups to stop in.

3. Setting STATEMENT COUNT on first TEST panel.
4. Specify "Select statements to stop on = yes" ahead of time.

Selecting Statements to Stop On

Choosing to stop at particular statements is often referred to as setting STOPs or BREAKPOINTs. After answering YES to "Select Statements to Stop In," the screen shown in Figure 4.18 displays all processes and groups for the current application.

Enter either an "S" or an "L" before the process/group names you are interested in. Place an "S" by a group/process to stop before executing its first line. Place an "L" by a group/process to display its statements (Figure 4.19). Key in a "C" for a group/process with STOPS SET=YES to turn off stops within its statements. If multiple applications will be tested (a CALL, XFER, or DXFR will be executed in this test), press PF4 to view groups and processes in the next application. Press PF3 to leave this procedure and continue testing.

After using the "L" to list the statements in a group or process, the screen in Figure 4.19 appears. Select the statements you wish to stop before executing. Place an "S" on the line you wish to stop on. You may also stop before executing a process option. And, you may stop on lines that are part of a FLOW (if you are using FLOWs). Stops are not carried across XFERs or DXFRs.

```
EZEM9L                         TEST
EZE00661A Enter an 'S' to stop before executing the statement


                        PF3 = Continue
             Application = applnam  Process or Group = TZ01030
Total lines 0007 .................. SELECT STATEMENTS TO STOP ON ............................................
***                       TOP OF LIST
001 MOVE TZ01M.KEY TO TZ01R.KEY;
002 ---INQUIRY TZINPUT EZERTN----
S   IF TZINPUT IS NRF;
004   PERFORM TZ01EOF;
005 ELSE;
006   PERFORM TZ01RP;
007 END;
```

Figure 4.19. Test—Select statements to stop on.

Sample of Trace Output

A sample of trace output is displayed in Figure 4.20 for your examination. The actual content of the trace output will be controlled by test selection criteria and the flow of the application being tested.

```
>>>> DOPCT;

              ----> 001.    TZTS010 ----> 002. DOPCT  ---->
>>>> TBSUB = RECID.FIELD1 + 12;

         TBSUB     OF      WKREC = 0244
>>>> WHILE TBSUB < 300;

         +0244   LT +0300 = YES
```

Figure 4.20. Test—Sample trace output.

Normally, the trace display shows CSP statements before they are executed; the actual statements are preceded by a series of "greater than" symbols (>>>>). After a branch, the path used to get to the branch is depicted graphically with an arrow (-->) pointing in the direction of the branch. Returns from branches are also shown, with the direction of the arrow reversed (<--). Unless turned off by the user, whenever a data item is modified, the new contents of the data item are displayed. After a comparison, the actual compared values and a YES/NO indicator are shown. This can get tricky until you get used to the format, for instance:

```
"NO" NE "YES" = YES
```

means that CSP has compared "NO" to see if it is not equal (NE) to "YES" and found the statement to be true (YES).

Commands Available at Stop Points

While you are stopped for one of the reasons discussed earlier, you have the ability to display or modify data items. To view the current contents of a data item, use the SHOW command. Subscripted items may also be shown; use either a constant or another CSP data item as the subscript (Figure 4.21).

You may alter data by entering a MOVE command (Figure 4.22). Any MOVE statement that would be valid within a process and/or group may be entered here. COBOL programmers, be careful. CSP does not know the family of COBOL figurative constants (SPACES, ZERO, and so on).

SHOW itemname

---

or

SHOW object.itemname

Figure 4.21.   SHOW command syntax.

> **MOVE newval TO itemname**

Figure 4.22. MOVE command syntax.

CONCLUSION OF TEST

When testing is finished, the panel illustrated in Figure 4.23 appears. The system provides a great deal of information concerning the test just completed. You may even restart the test. In fact, pressing ENTER does just that. New users, take a while to get used to the fact that you must press PF3 to finish the test when this panel appears.

```
EZEM96                          TEST

==>
                ENTER = Continue     PF3 = Exit
                   Application = applnam
.......................................... FINAL TEST STATUS ........................................


Warning messages generated      =   000
Last process or group executed  =   xxxxxx
Last statement executed:
      EZECLOS;

Restart test cycle              => YES
Restart record number           =>

Additional Error Messages:  Total Lines 0000
***                     TOP OF LIST
***                     END OF LIST
```

Figure 4.23. Test—Final test status.

CHAPTER 4 EXERCISES

Questions

1. CSP "programs" are called _____.
2. CSP modules are called _____.
3. Why is a naming convention necessary in CSP?
4. What are the rules for naming CSP routines?

5. When viewing the Application Process List, what does PF4 do?
6. What is the function of the semicolon in CSP statements?
7. What does EZECLOS do?
8. What is a CONVERSE process option? Where is it specified? How is it executed?
9. What should be specified at the beginning of every test?
10. Why is marking "ALLOW IMPLICITS => YES" sometimes a problem?

Computer Exercise

Building and Testing a First CSP Application

1. Define an application called BB01A that will do nothing but display the map created at the end of Chapter 3 (map group BB01G, map name BB01M01) one time. The application should end immediately after the user responds to the map.
2. Two processes should be defined. Use the examples in Figures 4.7 through 4.9 as models.
   Look at Figure 4.24.
   Name the mainline process BB01P-MAIN.
   Name the converse process BB01P-CONVERSE.

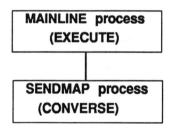

Sample pseudocode:

```
Begin
    Display single map
End
```

Figure 4.24. Sample structure chart and pseudocode.

3. Use the interactive TEST facility to execute your application.
4. Try setting stop points, using statement count, SHOW, and MOVE while testing.

CSP/AD Statements

As described in Chapter 4, like other programming tools, CSP provides statements to do the "nitty gritty" work of the application. These statements provide arithmetic, assignment of variables, conditional processing, and branching for CSP applications. It is important to realize that Input and Output (I-O) functions are *not* done with CSP statements. Instead of doing I-O with statements, CSP I-O is provided by process options and is integral to the process activity. I-O process options are descriptive of the work being done. For instance, CONVERSE is used to send and receive maps, and INQUIRY is used to read records. CSP statements are used to make decisions and process data *before* the I-O activity is attempted and *after* the I-O activity is complete. Together, CSP statements executed within processes are the programming capability of CSP. CSP processes may be executed by PERFORMing them (the preferred "structured" method) or by using a fall-through, go-to mechanism called FLOWs (against standards in most shops).

PROCESS OPTIONS

As discussed previously, CSP processes represent individual logic functions required to achieve the goals of the application.

COBOL programmers often remark that processes are logically similar to COBOL's paragraphs. The designers of CSP opted to have separate processes defined for different programming functions. The type of activity done by a particular process is defined by its associated PROCESS OPTION. While this takes some getting used to, once you have a good handle on the process mechanism and the different process options available, this setup will save you a great deal of time.

As an introduction, the different process types are listed by function below. Each process option is examined more fully later in this text.

Processing only, no I-O

> EXECUTE

File/Database I-O

> INQUIRY, UPDATE, REPLACE, ADD, DELETE,

> SETINQ, SETUPD, SCAN, SCANBACK, CLOSE, SQLEXEC

Map I-O

> DISPLAY, CONVERSE

This chapter discusses the EXECUTE process, STATEMENT GROUPs, and various CSP statements that may be used within processes. The other I-O-oriented process types are discussed in separate chapters depending upon their function later.

EXECUTE PROCESSES

The default process option on the Application Process List is EXECUTE (Figure 5.1). Enter a process name, place a "P" in the SEL column, and press the ENTER key to begin process definition. A screen similar to Figure 5.2 should appear next.

EXECUTE processes represent blocks of code not associated with any particular I-O function. All other processes do some kind of input or output processing besides the code provided by the statements.

```
EZEM36                    APPLICATION DEFINITION

==>
              ENTER = File and continue       PF3 = File and exit
              PF4 = Display application structure
                        Application Name = SMPL010
Select Definition:  S = P+F+L   P = Processing  F = Flow   E = Edit Object
              O = Object Selection          L = Structure List
Total lines 0001 .........................APPLICATION PROCESS LIST .....................................................
SEL PROCESS                   OPTION        OBJECT                ERROR
***                       TOP OF LIST
P01 my-mainline               EXECUTE
***                       END OF LIST
```

Figure 5.1.   EXECUTE process—Application process list.

```
EZEM39              APPLICATION PROCESS DEFINITION

==>
              PF3 = File and exit  (or file and continue if more selected)
       Process => MY-MAINLINE      Description =   Application mainline routine
       Option = EXECUTE       Object =
Total lines 0012 ...............................STATEMENT DEFINITION ...........................................................
***                       TOP OF LIST
***                       END OF LIST
```

Figure 5.2.   EXECUTE process—Statement definition.

Statements within EXECUTE processes are used to PER-
FORM other processes. No error routines or objects are needed
by EXECUTE processes. CSP also can create something called a
"Statement group." Statement groups are logically similar to EX-

ECUTE processes and have exactly the same performance. For reasons unknown to the author, CSP provides both of these almost identical tools but places restrictions on their use. EXECUTE processes and statement groups may be used interchangeably EXCEPT for a few particular cases. MAIN processes (those that appear on the first PROCESS LIST panel, on level 1) may only be actual CSP processes. Therefore, each application's "mainline" routine is usually an EXECUTE process. There are several places in CSP where only statement groups are allowed: the TEST statement, the FIND statement, error routines for record I-O processes, and as automatic edit routines. No explanation is provided by IBM for the need of two such constructs. Don't worry about it—just use them according to their limitations.

The Application Process Definition—Statement Definition panel (Figure 5.2) is where CSP statements are entered. Code on this panel represents a logical function in the application's design. At the top of the panel the process name and description are listed along with the process option specified. If a process option other than EXECUTE is specified, the object of the process's I-O will also be listed (more on this later). Statements must be ended by a semicolon (;) in CSP. This allows statements to be continued over several lines. Before CSP V3R3, CSP did not allow statements to be continued and would insert the semicolon automatically (it no longer does this). Any text following a semicolon is treated as a comment. It is common practice to insert lines beginning with semicolons for documentation and readability purposes.

STATEMENT GROUPS

CSP processes revolve around the selection and execution of PROCESS OPTIONs (usually involving some form of built-in I-O). CSP allows execution of either processes or statement groups. A statement group is a collection of statements without a process option, thus, not containing any I-O. A statement group may PERFORM an I-O process (with one exception, a statement group executed as a Field Edit may not perform I-O at all). There is little difference between a statement group and an EXECUTE process, both may include any CSP statement, and neither directly performs I-O. Only statement groups may be used for automatic field edits and TEST statements.

As I-O error routines, statement groups are treated in a branch-and-return fashion. EXECUTE processes named as I-O error routines must be "main" processes. They are branched to directly when an error occurs and application FLOW logic takes over. Statement groups are easier on an application's structure and do not violate most shops' "structured" programming rules.

Statement group definition is option 7 from the DEFINITION SELECTION screen. Or, when writing CSP process statements, the act of executing a statement group begins the definition process. Execution of processes and statement groups is discussed later in this chapter.

Using statement groups and EXECUTE processes, we can create routines that are available to multiple applications, allowing the sharing of complex and/or oft-repeated code. See Figures 5.3 and 5.4.

```
EZEM3R                     STATEMENT GROUP DEFINITION


==>
                    ENTER = Continue        PF3 = File and exit
                 Statement Group Name = xxxxxxxxxxx
.................................. .......................... FUNCTION SELECTION ............................................



                       Enter number of function desired

                       1   Statement Group Specification
                       2   Statement Definition
                       3   Structure List

```

    – **Enter this process by choosing option 7 "Statement Group" on the DEFINITION SELECTION panel (EZEM10)**

    – **This panel does not display when dynamically creating Statement Groups as part of coding process statements**

Figure 5.3.   Statement group definition.

```
┌─────────────────────────────────────────────────────────────────────────┐
│ EZEM3B                    STATEMENT GROUP DEFINITION                       │
│                                                                           │
│ ==>                                                                       │
│           PF3  =  File and exit  (or file and continue if more selected)  │
│           PF4  =  Format processing statements          PF5  =  Validate  │
│      Group = xxxxxxxxxx          Description =  Xxxxxxxxxx sample statement group │
│ Total lines 0012 ...............................STATEMENT DEFINITION .........................................│
│ ***                          TOP OF LIST                                  │
│ ***                          END OF LIST                                  │
│                                                                           │
│                                                                           │
│                                                                           │
│                                                                           │
│                                                                           │
└─────────────────────────────────────────────────────────────────────────┘
```

- This panel appears when you choose option 2 STATEMENT DEFINITION from the STATEMENT GROUP DEFINITION (EZEM3R) panel

- This panel also displays after using the "P" option from the APPLICATION PROCESS LIST (EZEM36) or STRUCTURE LIST (EZEM37) panels for a Statement Group (designated by the word GROUP in the OPTION column)

Figure 5.4.   Statement group—Statement definition.

CSP/AD STATEMENTS

CSP's processes may include many different types of statements. These statements provide a great deal of functionality common to other programming tools. CSP provides statements to do a variety of programming tasks including the following:

Arithmetic

add (+), subtract (–), multiply (*), divide (/), remainder (//)

Data manipulation / Assignment statements

MOVE, MOVEA, RETRieve, SET

Conditional processing / Looping

IF, WHILE, TEST, FIND

Logic Flow

> PERFORM, stmtgroup name, special EZE words

Application control

> CALL, DXFR, XFER

Flow logic

> EZEFLO

Most of CSP's statements are discussed in this chapter. The application control statements CALL, DXFR, and XFER as well as some special CICS CALLs will be covered in Chapters 13 and 18. The two table handling statements FIND and RETRIEVE have also been omitted from this chapter and will be discussed completely in Chapter 10.

ARITHMETIC

CSP arithmetic is similar to many other programming tools. A result field is named to store the product of a calculation. The basic arithmetic conditions are supported by CSP. CSP's arithmetic is excellent for most of a regular system's needs. But, CSP is not the best place for complex calculations and statistics. Other programs should be executed (see CALL processing in Chapter 13) when complicated calculations are required. Figure 5.5 illustrates the basic syntax of CSP arithmetic statements.

The result field of a calculation may be any variable known to the CSP application. It is common to qualify the name of a variable with the object (record/map) name in which it resides. For instance, WSRECORD.VALUE1 = 1 + 1; places the result of the addition statement into the VALUE1 field that is part of the

```
object.result = operand1  operator  operand2 ...
                       operator  operandn
             (r
```

Figure 5.5.   CSP arithmetic syntax.

object named WSRECORD. Field names that are unique to an application do not require qualification, but it is never wrong. Fully qualifying names makes application maintenance easier.

Operands in the expression may be numeric literals (e.g., 1, –10.5), data item (field) names, or special EZE function words.

Valid operators allowed by CSP are listed below:

Operator	Function	Example
+	add	mapnam.fld1 = mapnam.fld1 + 1;
–	subtract	recrd.fld = mapnam.fld2 – mapnam.fld3;
*	multiply	recrd.flda = mapnam.fld3 * .15;
/	divide	recrd.fldb = 100 / recrd.flda;
//	remainder	recrd.fldc = 100 // recrd.flda;
		(result is remainder after division, for example, A = 10 // 3; would place a value of 1 in A)

Multiple arithmetic operations may be combined (e.g., recrd.fldd = mapname.fld5 * 100 + 1;). If multiple operations are combined in one statement, multiplication and division are done first, followed by addition and subtraction. If priority is equal, calculations are done from left to right. Unfortunately, CSP does not provide the capability of influencing calculation order with parentheses (this is being addressed in Version 4).

All calculation results are truncated unless rounding is specifically requested using the "(r" option. By placing an open parenthesis followed by an "r" you may request that CSP round the results of a calculation. CSP rounds using the standard rule of rounding up at .5 or greater and truncating if less than .5. It is important to understand that rounding is performed on the final result only; no intermediate rounding occurs.

DATA MANIPULATION/ASSIGNMENT STATEMENTS

The MOVE statement (Figure 5.6) is used to replace the contents of a field entirely with another field's value or a literal. MOVE copies the contents of operand1 into operand2; operand1 remains

---

MOVE operand1 TO operand2

Figure 5.6. MOVE statement syntax.

unchanged. Operand1 may be a literal, data item, record, map, subscripted data item, or one of several CSP EZE names (see the CSP/AD Developing Applications manual (V3 R3) or the CSP/AD Operation—Development manual).

The keyword TO is optional (but it seems like a very good idea from the readability standpoint). Operand2 may be a data item, subscripted data item, record, map, or one of several CSP EZE names.

Moving a map name or record name to another map or record causes all like-named fields to be moved (similar to COBOL MOVE CORRESPONDING). This may seem like a good idea, but it may lead to maintenance difficulties later. MOVE should be data-type consistent for efficiency reasons. CSP will attempt to provide data conversions, but the conversions may be expensive performance-wise. Numeric data should only be moved to numeric fields (CSP will provide conversion). Numeric moves are always decimal-aligned by CSP based upon field definitions.

Character data should only be moved to character fields. Character to numeric (and vice versa) moves will be attempted. CSP will first validate the data to make sure it is numeric (and without decimals), and then CSP will attempt to convert data. Negative numbers may convert invalidly from character.

Conversion is performed for CHA to HEX moves automatically. Moving a longer character field to a shorter field results in truncation to the shorter field's length. Moving a shorter character field to a longer field causes the shorter field's data to be padded with blanks. Don't forget, the TO is optional but still a good idea.

CSP provides a statement specifically for moving data from one array to another called MOVEA (Figure 5.7). MOVEA is either a perfect monstrosity or a perfect tool, depending upon your orientation. MOVEA allows the movement of data from/to selected rows in an array.

MOVEA operand1 TO array2 FOR operand3

Figure 5.7. MOVEA statement syntax.

Operand1 may name any literal, data item, array name, subscripted data item, subscripted array, or several EZE fields. Operand1 describes data to be moved or array item to begin with.

Array2 names the target array. Array2 may be subscripted and describes the target array or target array and beginning array item to move to.

Operand3 specifies the number of array items to be moved from inoperand to outoperand (limited by the size of outoperand). Operand3 may be a literal, data item, or EZE word. Its value must be between 0 and 64,536 (inclusive). Data items referenced in MOVEA may be subscripted.

A few examples of MOVEA and its use are shown in Figure 5.8. If operand1 is an array and operand3 is left out, operand3 defaults to the smaller of operand1 and array2. If operand1 is not an array and operand3 is left out, operand3 defaults to (result array size—subscript + 1). After MOVEA is complete, the CSP special name EZETST will contain the subscript of the last element in the result table changed by this operation.

MOVEA follows the same rules as the MOVE statement for data conversion, and once again, TO is optional. While MOVEA certainly provides a great deal of functionality, its use should be well documented. Future application maintenance people should be told to clearly state what the *intent* of the MOVEA statement is.

CSP provides the SET command with many functions to ease programming tasks. SET is used to clear a display or record, change an attribute, position a file, position the cursor, or position a printer. Several examples are provided here to illustrate the different syntax uses.

Used when processing INDEXED records, SET recordname SCAN (Figure 5.9) locates a record in the file associated with recordname that matches the current key field value. This is similar to the START verb used in COBOL, but in CSP no I-O return code results (more on this later in Chapter 11).

To initialize a record definition's fields, the SET recordname EMPTY command is used (Figure 5.10). EMPTY sets all numeric fields in the record to zero (0), all character/mixed/dbcs fields to blanks (' '), and all hex/binary fields to binary zeroes (x'00'). It is a good idea to use SET recordname EMPTY at the beginning of

```
MOVEA 'x' TO OUTARRAY;

    OUTARRAY (before)        0 1 2 3 4
    OUTARRAY (after)         x x x x x
```

```
MOVEA 'x' TO OUTARRAY(2) FOR 2;

    OUTARRAY (before)        0 1 2 3 4
    OUTARRAY (after)         0 x x 3 4
```

```
MOVEA INARRAY(3) TO OUTARRAY(2);
    INARRAY                  a b c d e
    OUTARRAY (before)        0 1 2 3 4
    OUTARRAY (after)         0 c d e 4
```

```
MOVEA INARRAY(3) TO OUTARRAY(2) FOR 2;
    INARRAY                  a b c d e
    OUTARRAY (before)        0 1 2 3 4
    OUTARRAY (after)         0 c d 3 4
```

• **NOTE: EZETST will contain the subscript of the last value changed by the command**

Figure 5.8.   MOVEA examples.

```
SET  recordname  SCAN
```

Figure 5.9.   SET statement—File positioning.

```
SET  recordname  EMPTY
```

Figure 5.10.   SET statement—Record initialization.

an application's execution, especially if new records are being created.

Other languages (COBOL, PL/I) require NULL indicators when processing SQL data; CSP does not. The command in Figure 5.11 sets sqlrec.item to a default value and turns on its NULL indicator. This is used to modify SQL database data to NULL (the SQL columns must be defined as NULL-capable in the database).

SET SQL sqlrec.item   NULL

Figure 5.11.   SET statement—SQL NULLs.

The command in Figure 5.12 is used for control over an entire map. The PAGE operand clears the screen before the next display or sets the printer to the top of the page. PAGE takes effect only when a CONVERSE or DISPLAY process is executed (CSP performs an automatic SET PAGE any time a new fixed map is being sent to the same terminal previously occupied by a fixed map). ALARM causes the terminal alarm to sound on the next CONVERSE (generally considered user *un*friendly). CLEAR resets a map as it was originally defined (not tied to a DISPLAY or CONVERSE). It is usually a good idea to SET mapname CLEAR at the beginning of an application using the map. EMPTY sets contents of numeric fields to zero and character fields to blanks on the map. EMPTY's initialization of fields should be anticipated by program design and may increase communications line traffic (adding expense).

CSP provides the ability to control several aspects concerning map fields, but some options are exclusive of each other. Figure

---

SET mapname   PAGE
ALARM
CLEAR or EMPTY

Figure 5.12.   SET statement—Map control.

```
SET mapname.item NORMAL or DEFINED
                 CURSOR
                 FULL
```

```
SET mapname.item CURSOR
                 FULL
                 MODIFIED
                 BRIGHT or DARK
                 PROTECT or AUTOSKIP
```

Figure 5.13.   SET statement—Map data item control.

5.13 illustrates the allowable combinations. Setting a map field to NORMAL causes the field's attribute to be set to normal intensity, unprotected, and unmodified. DEFINED sets a field's attribute to the value defined at map creation.

CURSOR places the cursor on the first character of this field the next time the map is DISPLAYed or CONVERSEd (if more than one has been SET, the last wins).

FULL is used to control the field's display if the field is empty on the screen (blank). If the field is empty, FULL fills it with asterisks prior to display. Any asterisks remaining in a field after entry of "real" data must be cleared by the user (either overkeyed or use ERASE EOF) or they will be brought back into CSP. This will require the application to edit them out.

MODIFIED "tricks" the terminal into thinking that data for this field has been changed (turns MDT "on").

BRIGHT sets bright intensity for a field; DARK sets dark intensity for a field (not usually a good idea but useful for security purposes sometimes).

PROTECT protects a field from user entry. If the field in front of this one is unprotected, the cursor will stop and the keyboard will lock if the user tries to enter data here (the user should press the tab key to move on).

AUTOSKIP protects a field also, but it causes the cursor to skip automatically to the next unprotected field (usually considered the most user-friendly).

Please notice that CSP does not allow control of field color with the SET statement other than default colors applied by the hardware based upon intensity and protection:

protected/autoskip + bright defaults to white,

protected/autoskip + normal defaults to blue,

unprotected + bright defaults to red,

unprotected + normal defaults to green.

Version 4 will allow explicit SETting of color.

Conditional Processing/Looping

CSP provides conditional processing with four different statements.

IF   (standard if-then-else construct)

WHILE   (do-while looping mechanism)

TEST   (specialized if-then-else construct)

FIND   (if-then-else based upon table/array search)

The IF and WHILE statements allow specification of a condition to be tested. The conditional tests used by both IF and WHILE statements are the same, so they are covered together here. TEST takes one action if a specifically defined condition is true, and another if it is false. FIND takes one action if a value is found in a table/array and another if it is not (tables and table-related statements like FIND are covered in Chapter 10). The syntax of IF and WHILE statements is shown in Figure 5.14.

```
IF condition1                       ; this is a
[AND / OR      condition2...]        ; good
     resultoption1                  ; way to include
ELSE                                ; comments
     resultoption2                  ; in the
END                                 ; code...
```

Figure 5.14.   IF-THEN-ELSE statement syntax.

IF does the resultoption1 statement(s) if the condition(s) are true. ELSE causes resultoption2 statement(s) to be executed if the IF condition(s) are false. IF, ELSE, and END must be coded on separate lines. ELSE is optional. Either result option may be null. END is required for each IF. Be careful to code semicolons (;) everywhere they are required.

WHILE (Figure 5.15) causes a loop executing the resultoption statement(s) as long as the condition(s) are true. It is possible not to execute the resultoption statement(s) at all if the condition(s) are false the first time the statement is executed (this is a do-while).

```
WHILE condition1;
[AND / OR condition2...];
    resultoption;
END;
```

Figure 5.15.   WHILE statement syntax.

END is required for each WHILE. Most applications code a WHILE loop in the "mainline" process that ends only when the user signals the desire to quit. Inside the WHILE loop all map display processes are performed so that the WHILE condition is tested between the user response and the next display of the map. Tests for function keys (PF3 and PF12 are SAA keys used to end and exit respectively) or for so-called "quit" values are coded in the WHILE loop. Using a WHILE in this fashion allows clear control of the user session. For both IF and WHILE statements, AND/OR must begin on a new line (AND and OR are mutually exclusive until Version 4). Parentheses may *not* be used to logically group tests.

Conditions in IF and WHILE statements may use any of the following comparators in a comparison statement (item1 comparator item2).

Comparators	How Test Works
EQ or =	item1 equals item2
NE or ¬= or =¬	item1 not equal to item2

GT or >   item1 greater than item2

GE or >= or =>   item1 greater than or equal to item2

LT or <   item1 less than item2

LE or <=   item1 less than or equal to item2

IN   tests if item1 is a value in the array item2, if item2 is
                            subscripted. IN looks for a match of item1 from that
                            subscript on in the array. EZETST will be set to zero if
                            no match occurs and to the subscript of the first
                            match if a match is found.

IS   item1 compared to condition identified by item2 (see
                            TEST conditions on following pages)

NOT   opposite of IS, true if item1 does not match the
                            condition specified by item2

AND   used to logically tie 2 or more conditions; if all are
                            true, the statement is true

OR   used to logically tie 2 or more conditions; if one is
                            true, the statement is true

AND may not be used in the same statement as OR (this is being fixed in Version 4). Data items in the comparisons may be literals, data items, or one of several CSP EZE words (see Appendix C at end of book). Result options may specify one or more statements, the names of statement groups (discussed later), or perform CSP process options.

USING "IF" OR "WHILE" WITH PREDEFINED CONDITIONS

CSP allows IF and WHILE statements to test for several predefined conditions. These predefined conditions may also be used in conjunction with the TEST statement (covered later in this chapter). The syntax is specific—"IF xxx IS condition" or "IF xxx NOT condition"—the words IS and NOT may not be used together.

```
IF mapnam IS/NOT MODIFIED
```

Figure 5.16.   Map modification test.

To see if a map has been modified by the user, use the statement in Figure 5.16. IS MODIFIED shows that data has been entered on the map this time; NOT MODIFIED indicates that the map is unchanged.

To test a map field, use the statement in Figure 5.17. In this statement BLANK, BLANKS, NULL, and NULLS all mean the same thing: that a field contains blanks and are interchangeable. These tests check to see if the data received from the panel contains all blanks or nulls (caused by pressing ERASE EOF on a 3270, gets converted to blanks). BLANK/NULL may also indicate that no data has been moved into the field and the contents are as originally defined. DATA is the opposite of BLANK or NULL. CURSOR tests to see if the cursor was positioned on this field when the user responded. MODIFIED checks to see if data in the fields was changed. Setting of the MDT (Modified Data Tag) via the SET . . . MODIFIED command or if the MDT was set on as part of map definition will cause the test to be true.

A simple test may be made to detect if a specific function key has or has not been used. This test makes use of the special CSP variable EZEAID. EZEAID is a one-character field containing a value showing which button was pressed when the user responded to the map (see Figure 5.18). All normal function keys may be tested for including PF keys, PA keys, and ENTER. CICS users should note that there is no way to test for the CLEAR key

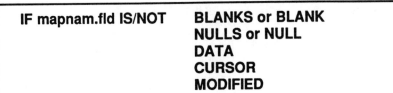

Figure 5.17.   Map data item tests.

IF EZEAID IS/NOT	**PF1-PF24 or PF**
	PA1-PA3 or PA
	ENTER
	BYPASS

Figure 5.18.   Function key tests.

(a normal CICS programming technique). The CLEAR key is not supported since its function is not used in all CSP environments.

PF	checks if any PF key was pressed
PF1-PF24	checks if the specified PF key was pressed
PA	checks if any PA key was pressed
PA1-PA3	checks if the specified PA key was pressed
ENTER	true if the ENTER key was pressed
BYPASS	tests true if the function key pressed was on the list of BYPASS keys defined for the MAP or for the APPLICATION

SQL data items have two special tests for NULL data and for TRUNCation. Unlike other languages, CSP does not require special indicator variables to process NULL SQL columns. The ability to test for NULL and SET to NULL make SQL's use of NULL data much simpler in CSP. TRUNC indicates that the data field in CSP is too small to hold the entire SQL column's data, and that data has been truncated (see Figure 5.19).

IF sqlrec.fld	**IS/NOT**	**NULL or TRUNC**

Figure 5.19.   SQL NULL/TRUNC test.

CSP provides a couple of different methods for checking the success/failure of record I-O. The simple tests allowed are consistent for all types of record definitions supported by CSP (see Figure 5.20). However, some people find it more meaningful to

IF recnam	IS/NOT	DED or DUP or EOF or ERR or FUL or HRD or LOK or NRF or UNQ

Figure 5.20.　Record I/O result testing.

check for actual response or SQL codes instead. The chapters on data I-O and SQL processing list other tests that may be used.

DED　　　　　　A deadlock occurred, the I-O failed (SQL only).

DUP　　　　　　A duplicate key existed; the I-O function still works in environments allowing duplicate keys (e.g., VSAM AIX).

EOF　　　　　　End of file

ERR　　　　　　Non-zero return code

FUL　　　　　　Disk area is full, hardware error only returned to program if the EZEFEC field is equal to 1.

HRD　　　　　　Hardware error, hardware error only returned to program if the EZEFEC field is equal to 1.

LOK　　　　　　File lockout, I-O fails (AS/400 or DPPX only)

NRF　　　　　　No record found by INQUIRY or UPDATE process, indicates logical end-of-data for SCAN or SCANBACK processes. When processing RELATIVE files indicates that the record position is empty.

You may have noticed a glaring omission. There is not a test to make sure that the I-O is okay! As much as most of us dislike negative tests, the best bet is "IF recname NOT ERR" to see if an I-O worked as designed.

All of the above conditions may be replaced with tests of the EZERT8 variable (eight-character field containing numeric data) for INDEXED, SERIAL, and RELATIVE files. SQL tests may be replaced with tests of the EZESQCOD variable. Appropriate values to test for may be found in the CSP "Messages, Codes, and Problem Determination" manual, or in SQL references.

TEST operand condition trueoption,falseoption

Figure 5.21.   TEST statement syntax.

Test Statement

Many of the special condition tests discussed for IF and WHILE may be replaced by the TEST command. TEST (Figure 5.21) provides a shorthand method for testing IF-THEN-ELSE logic that some find quite appealing. Beware, others find TEST less meaningful and comments should probably be used to record your intent. TEST compares the named operand to the named condition. If the named condition is true, the trueoption (if specified) executes. If the named condition is false, the falseoption is executed. Either the trueoption or the falseoption may be omitted (null THEN or ELSE).

Operands tested may be a map, map item, record, data item, or EZEAID (EZEAID indicates which function key the user pressed). Condition may be many things depending upon the operand specified (all TEST statements may be replaced by IFs). If the operand is a map, TEST mapname MODIFIED is true if ANY field on the screen has an MDT turned "on." If the operand is a map item, the following operands may be used (only one per TEST though):

BLANKS or BLANK	Is data in field all blanks or nulls?
NULLS or NULL	Same as BLANK
CURSOR	Indicates if the cursor was positioned on a field at the time the function key was
DATA	Does data contain anything other than blanks or nulls?
MODIFIED	Was MDT on for this field?
n	Where n is a number that is equal to the number of characters returned in the field
+n	Where n is a number and the number of characters returned in the field is greater than or equal to n

–n

Where n is a number and the number of characters returned in the field is less than or equal to n

If the operand is EZEAID, the following conditions are allowed:

PF1-PF24

Tests a particular PF key value

PF

Tests if ANY PF key was pressed

PA1-PA3

Tests a particular PA key value

PA

Tests if ANY PA key was pressed

ENTER

Tests if the ENTER key was pressed

BYPASS

Was a BYPASS key pressed (BYPASS keys are defined as part of MAP and/or APPLICATION definition)?

If the operand is a SQL record item:

NULL

Indicates column is NULL on the database

TRUNC

Data was truncated on read

If the operand is a record in a serial or indexed file, you may test the results of the last I-O (actual system return codes are stored in the CSP special variable EZERT8).

DED

On DPPX indicates that a deadlock occurred on an indexed or relative record

DUP

Attempted to add a duplicate record; if duplicate keys are allowed in file the record was added.

EOF

Attempt to read past end of file (SCAN or SCANBACK)

ERR

Return code was not zero

FUL

Data set has run out of space

HRD

Hardware error on I-O

LOK

Valid on OS/400 and DPPX systems only;

	indicates lockout has occurred between two transactions
NRF	No record found on INQUIRY or UPDATE or SCANBACK; executed without setting valid key value
UNQ	Add failed because an alternate index file key was duplicated and duplicates are not allowed in the alternate file

If the operand is a record definition for a SQL database, values are set based upon the SQLCODE from the last I-O (SQLCODE is stored in the special CSP variable EZESQCOD).

DED	Deadlock, I-O failed due to contention
ERR	SQLCODE was not zero
HRD	SQLCODE < 0
NRF	SQLCODE = 100, no record found or end of cursor data
UNQ	ADD or REPLACE failed due to duplicate value in index column

For DL/I records see the CSP/AD Reference and the appropriate IMS or DL/I manuals.

Figure 5.22 shows the TEST statement being used to see if data was entered by the user (actually, were any MDT's "on" on the map). If any data was entered on the named map, the edit_process statement group is executed. If no data was entered, the empty_process statement group is executed. Only statement group names may be used as TEST statement result options. To refresh your memory, statement groups are logically equal to EXECUTE processes.

TEST mapname MODIFIED edit_process,empty_process;

Figure 5.22. Map modification test—TEST statement.

```
IF mapname IS MODIFIED;
    edit_process;
ELSE;
    empty_process;
```

Figure 5.23.   Map modification test—IF statement.

All uses of the TEST statement can be made using IF tests. The TEST in Figure 5.22 may be replaced by the IF statement in Figure 5.23.

Many people find the IF test more "pleasing to the eye" and self-documenting. Others like the reduced coding effort allowed by using TEST. If either is allowed by standards at your shop, I suggest that you experiment to see which fits your needs best.

LOGIC FLOW—UNCONDITIONAL LOGIC

It is possible to branch-to/execute statement groups or processes directly by using the available unconditional processing capabilities of CSP. To branch to a statement group and then return, simply enter the name of the group where a statement normally goes. To branch to a process and then return, use the PERFORM statement to PERFORM process-name. Figure 5.24 shows an EXECUTE process that executes either a process or statement group depending upon function key tests.

In a noneducational environment, it is best to standardize the use of EXECUTE processes and statement groups. Be aware that it is impossible to outlaw either one without losing functionality. In the cases above, the groups/processes are invoked as if they were coded in-line at that point. When the group/process completes its task, the next statement is executed. This BRANCH-AND-RETURN type of processing should be familiar to COBOL programmers used to PERFORMing paragraphs or sections, and to PL/I programmers used to CALLing procedures. The use of PERFORM signals CSP that a process is to be executed. The use of any word (accidentally or on purpose!) without a known CSP statement causes CSP to *assume* that the word requests execution of a statement group by the same name. If you exit from the

```
┌────────────────────────────────────────────────────────────────────────────
│ EZEM39                          APPLICATION PROCESS DEFINITION
│
│ ==>
│              PF3 = File and exit  (or file and continue if more selected)
│        Process = blahblahblah              Description = very interesting documentation
│        Option = EXECUTE                    Object =
│ Total lines 0015 .............................STATEMENT DEFINITION .....................................
│ ***                          TOP OF LIST
│  001 IF EZEAID IS BYPASS         ;   this is a test for the bypass key
│  002    IF EZEAID IS PF3         ;   if bypassing because of PF3
│  003       EZERTN               ;      exit & return
│  004    END                      ;      ends inside if
│  005    IF EZEAID IS PF12        ;      if bypassing because of PF12
│  006       EZECLOS              ;      exit
│  007    END                      ;          ends 2nd inside if
│  008 END                         ;   ends 1st if
│  009 IF EZEAID IS PF8            ;   if PF8 pressed
│  010    PERFORM scroll_fwd_routine;   scroll_fwd_routine is a PROCESS
│  011 ELSE;
│  012    IF EZEAID IS PF7         ;   if PF7 pressed
│  013       scroll_bwd_routine    ;      scroll_bwd_routine is a
│  014    ELSE                     ;          STATEMENT GROUP
│  015       PERFORM nothing_entered;
│
└────────────────────────────────────────────────────────────────────────────
```

Figure 5.24. Nested IF example.

process/group definition without deleting this, CSP will build an empty statement group in your MSL for you. It is a good idea to clean up accidental "orphans" as soon as they occur (otherwise, pack-rats that we are, they tend to remain forever).

Though I will attempt to avoid lengthy discussions on using the FLOW options (the world of GO TO code should be left behind), one oddity needs to be explained. To branch to a process from the FLOW section, simply enter the name of the process to be performed. Yes, this seems a bit inconsistent, but I think the developers worked hard to build in a GO TO function without using the GO TO verb. Use EZERTN to cause the execution path to immediately return to an invoking process or group.

FLOW LOGIC

Part of defining a MAIN process (those processes showing up on the APPLICATION PROCESS DEFINITION list when first presented) is the opportunity to add statements to the FLOW section. Most installations use only one MAIN process that causes all other processes and statement groups to be executed in a hierarchical fashion. Using the FLOW section allows use of fall through, go to logic in the application. In this age of "structured" programming techniques, FLOW sections seem anachronistic. Indeed, many shops have legislated against the use of FLOW logic altogether (probably a good standard for new development). But, (there's always a "but," isn't there?) many existing systems, especially older ones, use FLOW extensively, and you should be aware of how it works.

First, FLOW logic comes from the days before CSP was capable of branch and return logic. This left only the capability to "fall through" executing logic serially and to "go to" other routines. By separating the control logic like this, the original CSP design makes maintenance a little easier because you know where most of the branching code is stored. Unfortunately, like most direct branching logic, it is possible to branch incorrectly. Earlier in this chapter it was suggested that a good way to keep an online user looping through an application was to code a WHILE in the "mainline" process until they ask to leave. Some shops allow one use of FLOW logic with the "mainline" process's FLOW section naming the "mainline" routine itself. The FLOW technique simplifies the "mainline" logic by removing the loop. Be careful! This method causes infinite looping and it is a good idea to have an EZECLOS conditionally executed to allow you to quit.

The EZEFLO statement is used (like a statement group) to "go to" the first FLOW statement associated with the current MAIN process (against standards in most shops). In a FLOW routine, placing the NAME of a MAIN process on a line by itself (as if executing a statement group normally) causes CSP to GO TO the top of that routine. PERFORMed processes do not have FLOW sections; only those processes at the highest level (MAIN) may have FLOW sections. Again, most shops have outlawed the use of FLOW and EZEFLO for standardization reasons.

CSP SPECIAL "EZE" WORDS

CSP provides many special words beginning with the three characters "EZE." Some EZE words are used as statement groups or processes, while other EZE words are used as variables. Finally, some EZE words are used as parameters by CSP to control processing. Sadly, there is no way to discern from an EZE variable's name whether it is to be treated as a variable or as a routine.

A list of some of the more useful EZE words and their functions begins here. For the most complete information, consult the CSP/AD manuals.

EZE Word	Description
EZEAID	One character field containing a value that represents the function key pressed by the user
EZEAPP	Eight-byte character field that contains an application name. This name is used by DXFR and XFER to dynamically control the transfer-to name.
EZECLOS	Routine that CLOSES a CSP/AD application (default for last process in application)
EZECNVCM	Move a "1" to this one-byte field to cause EZECOMIT to execute automatically after every display caused by a CONVERSE
EZECOMIT	Routine that causes SQL COMMIT WORK under TSO or VM, SYNCPOINT under CICS, CHKP under DL/I, and DTMS COMMIT under DPPX
EZEDAY	Five-byte field containing the Julian date "YYDDD"
EZEDTE	Six-byte field containing the Gregorian date in the form "YYMMDD"
EZEFEC	Move a "1" to this field in SQL applications. If this one-byte field's value is "0" (the default) the application is terminated by hard I-O (and SQL) errors. If the value is "1," the application continues to process and is responsible for handling the error situation.
EZEFLO	Routine that causes current statement group or sub process to GO TO the FLOW portion of the main process
EZEMNO	Two-byte binary field used to obtain messages from the message file (more on this in Chapter 9)

EZEMSG	Field containing error message
EZEOVERS	One-byte numeric field set to "1" when arithmetic overflow occurs
EZERCODE	Four-byte binary field used to set a JCL return code
EZEROLLB	Routine that causes SQL ROLLBACK under TSO or VM, SYNCPOINT ROLLBACK under CICS, ROLB under DL/I, and DTMS RESET under DPPX
EZERT8	Eight-character field containing file status values after I-O
EZERTN	Routine that exits from current statement group or process; equivalent to EZEFLO if executed from main process.
EZESEGM	One-byte switch used to turn segmentation "on" or "off" (more on this later)
EZESEGTR	Eight-byte field containing the CICS transaction ID to be activated when returning from the next CONVERSE. This field is meaningful only when segmentation is in effect and useful only when DB2 is involved.
EZESQCOD	Contains the four-byte binary SQLCODE. Value 0 means everything is okay; positive values are warnings/informatory, negative values represent errors. (0 = okay, 100 = nrf/enddata, xx0 = error)
EZESQRD3	Contains SQLERRD(3), a four-byte binary field containing the number of rows processed by SQL UPDATE or DELETE
EZESQRRM	Seventy-character field containing text for the current SQL error
EZETIM	Eight-byte field containing current time in the form "HH:MM:SS"
EZETST	Two-byte binary field modified by most statements that manipulate or interrogate arrays

EZEUSR	Eight-byte field (ten on OS/400) contains different information depending on the environment:

TSO	User ID
CICS	Terminal ID
CMS	Logon ID
DPPX	User ID
OS/400	Job User Profile Name

CHAPTER 5 EXERCISES

Questions

1. What command is used to control looping in CSP?
2. What special rules pertain to the use of AND and OR in conditional statements?
3. Are semicolons required?
4. How are comments usually embedded in CSP statements?
5. What command(s) may be used to see if a map field named "CUSTNAM" has been modified?
6. How does CSP perform I-O?
7. What are the differences between EXECUTE processes and statement groups?
8. What command(s) may be used to see if the PF11 key was pressed?
9. Besides a condition to test, what is a required part of every IF or WHILE statement?
10. What statement means "exit current routine," return to higher level routine?
11. What statement means "exit application"?

Computer Exercise
Map Edit Process

In the previous exercises, you have created a map and an application to display the map. In this exercise you will modify the existing application to do some edits, looping through the application until the user is ready to quit. This exercise is broken into several steps. If you are new to CSP, do yourself a favor and do them one at a time. Don't forget to test before you continue.

1. Modify your application so that any time the user presses PF12, the application ends immediately. This logic is best placed at the beginning of the program and/or any loop within the program.
2. Modify your application so that if the user enters "QUIT" in the Inventory ID field, the application ends immediately. This logic is also best placed at the beginning of the program and/or any loop within the program.
3. Change the "mainline" routine so that execution of the CONVERSE process occurs inside of a WHILE loop. Use the conditions in items 1 and 2 to control when the loop stops.
4. Add logic to see if the INVENTORY ID field was entered by the user. If so, it must be "QUIT" or have a value between '000000000AAA' and '999999999999' (inclusive). A good way to do this edit is by PERFORMING a new EXECUTE process inside the "mainline" routine's loop immediately *after* performing the CONVERSE process. This makes sure that your design is modular and that the edit only occurs after the screen has been received.

   If the INVENTORY ID entered is "QUIT," exit the application.

   If the INVENTORY ID entered is valid (following the above rules), move a message "Entered ID OK" to the EZEMSG field before redisplaying the screen.

   If the INVENTORY ID entered is not valid, make the field bright intensity, position the cursor on the field, and move a message "INVALID ID, TRY AGAIN" to the EZEMSG field before redisplaying the screen.

   See Figure 5.25.

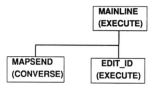

Possible pseudocode:

	PROCESS
Clear map & screen areas	MAINLINE
Move message to map asking for INVENTORY ID	MAINLINE
Do while user has not pressed PF3 or entered "quit"	MAINLINE
Display map	MAINLINE
Perform Edit INVENTORY ID	EDIT_ID
If INVENTORY ID OK	EDIT_ID
Move success message to map asking for next INVENTORY ID	EDIT_ID
Otherwise (INVENTORY ID failed edit)	EDIT_ID
Move error message to map	EDIT_ID
End Do	MAINLINE
Exit application	MAINLINE

Figure 5.25. Sample structure chart and pseudocode.

6

Data Definition

Up to this point, you have learned how to define maps, define applications, and create processes to manipulate the maps. In this chapter, you will learn how to define data records to CSP. While this chapter concentrates on non-database records, the same general principles covered here apply to database records (see Chapter 12). CSP works with a variety of data in each operating environment. All record types used must comply with restrictions of the testing and execution environments.

CSP/AD allows data records to be defined once and then reused. Further uses of the same data may be accomplished by simply specifying the already defined name as the object of an I-O process. CSP works with most types of data organization methods, but some types are not supported in all environments. Specifically, CSP provides the capability of defining data as seven basic types of records:

Indexed
Relative
Serial
Working-Storage
Redefined Record
DL/I Segment
SQL Row

Indexed implies that the data may be retrieved using a key of some kind. VSAM KSDS (Key Sequenced Data Sets) are used in MVS, VSE, and VM systems where the indexed record type is necessary.

Relative implies that the record is stored and retrieved using its physical location in the file (relative record number). VSAM RRDS (Relative Record Data Sets) are used in MVS, VSE, and VM systems where the relative record type is necessary.

Serial describes a simple sequential file. Depending upon the execution environment, any BSAM (Basic Sequential Access Method), QSAM (Queued Sequential Access Method), or VSAM ESDS (Entry Sequenced Data Set) may be described to CSP. Be careful! The CICS environment recognizes only VSAM ESDS as sequential files. If you need to access a regular sequential file (BSAM or QSAM) it must be defined as a Transient Data Queue.

CSP applications, like other types of programs, often require work areas for switches, counters, and other variables. By defining a Working-Storage record, internal variables used by the application may be described. Each application assumes that *one* Working-Storage record will be associated with it. The name of the appropriate Working-Storage record is entered as part of the Application Definition process. It is also possible to name other Working-Storage records in the Table and Additional Records List for an application.

Working-Storage structures are used to pass parameters back and forth when transferring control (calling) between applications. Sometimes a file will contain records of different descriptions (record type a, record type b). CSP allows definition of a Redefined Record to provide alternate data descriptions for already defined records.

DL/I Segments may be defined to CSP using the DL/I Segment definition. Prior to Segment definition, the PSB to be used must be defined. The PSB definition will obtain available database, segment, parent, and index information to CSP.

SQL Row records describe the RESULT of a query. The query used to create an SQL Row record may be from a table or view, and may represent a join of multiple tables/views. The usual SQL limitations on updates and insertions applies to these SQL Row definitions.

As with map and application definition, CSP records are defined using a series of panels. The first time a record is defined, CSP/AD will "walk" you from one panel to the next as the record is defined. After that, a menu-style panel lets you choose which portion of the record definition to view or modify.

Beginning in Chapter 8, you will learn to use the record definitions with CSP's I-O processes. It is usually easiest to define records before beginning coding of an application. Like all CSP objects, record definitions need to be done only once. All application developers having access to the MSL may use a record for application development.

DATA DEFINITION

Data records are defined using the following series of screens:

DEFINITION SELECTION
RECORD DEFINITION FUNCTION SELECTION
RECORD DEFINITION RECORD SPECIFICATION
RECORD DEFINITION DATA ITEM DEFINITION
RECORD PROLOGUE

DEFINITION SELECTION is the panel used to enter the record definition process. RECORD DEFINITION FUNCTION SELECTION is the panel displayed every time record definition is begun (except for the initial creation of the record when it does not display). RECORD DEFINITION RECORD SPECIFICATION is used to describe the type of record, connection to system data, keys (if applicable), and other information. RECORD DEFINITION DATA ITEM DEFINITION provides the capability to define/modify individual fields. RECORD PROLOGUE is a built-in place for documenting use and any special information concerning the record.

RECORD DEFINITION

To begin the record definition process, choose option 2 (Definition) from CSP's "Main Menu," the FACILITY SELECTION panel, or enter "=Edit" on the command line (see Figure 6.1).

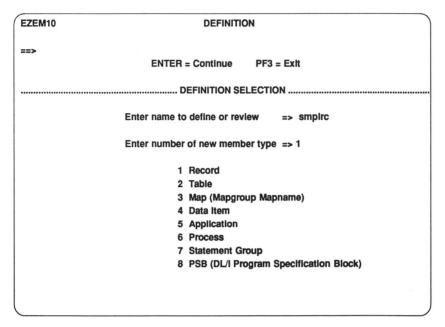

Figure 6.1. Definition selection.

Enter the name of the record to be created/edited in the position provided, and enter "1" as the new member type. If the record definition already exists in one of the MSLs in the current MSL list, the RECORD DEFINITION FUNCTION SELECTION panel (Figure 6.2) will appear next. If you are creating a new record, the RECORD DEFINITION RECORD SPECIFICATION panel (Figure 6.3) appears.

Record Definition—Function Selection

When editing an existing record definition, the panel in Figure 6.2 appears first. The four choices allow you to modify all aspects of the record's definition. Enter the number corresponding to the desired function:

Record Specification

Information about the entire record definition including record type, its connection to the system, key (if applicable), and default setting for variable scope

Data Item Definition

Add/modify field definitions

Record Prologue

Add/modify record documentation

Default Selection Conditions Definition

Modify the SQL statement (SQL only)

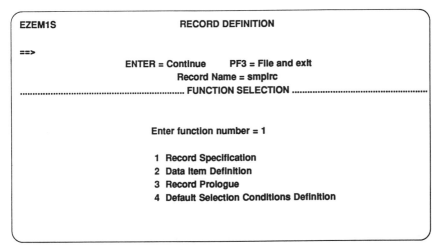

```
EZEM1S                    RECORD DEFINITION

==>
                ENTER = Continue      PF3 = File and exit
                       Record Name = smplrc
...................................... FUNCTION SELECTION ......................................


                   Enter function number = 1

                   1  Record Specification
                   2  Data Item Definition
                   3  Record Prologue
                   4  Default Selection Conditions Definition
```

Figure 6.2.   Record definition—Function selection.

Record Definition—Record Specification

The first time a record is defined, CSP automatically moves from panel to panel beginning with RECORD DEFINITION RECORD SPECIFICATION (Figure 6.3). Choose the record organization type from among the choices, then PRESS PF3 (oh consistency, where art thou?) to continue. During the initial definition, pressing PF3 moves you from one panel to another. There is no way to get CSP to cycle through these panels automatically once an object exists. Instead, you must return to the RECORD DEFINITION FUNCTION SELECTION panel (Figure 6.2) and proceed from there.

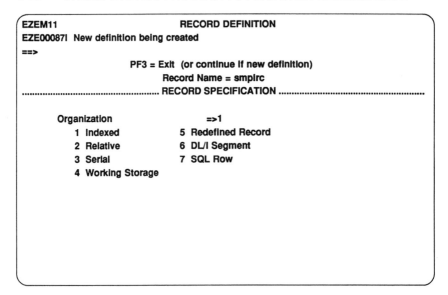

Figure 6.3. Record definition—Record specification (initial panel).

Choose the organization type that meets your needs:

Indexed Data retrieved using a key

Serial Data processed sequentially

Relative Data processed by relative record number

Working-Storage Work fields used by applications

Redefined Record Define additional record layout for existing record

DL/I Segment Record data described by DL/I segment

SQL Row Result of an SQL query

After choosing an organization type, CSP redisplays this panel with additional fields based upon the organization type selected. The next several pages show the different organization type screen presentations. Each panel used is similar to the others, with organization type specifics included.

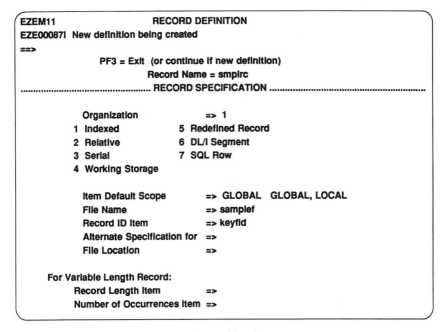

EZEM11   RECORD DEFINITION
EZE000871 New definition being created
==>
                    PF3 = Exit (or continue if new definition)
                        Record Name = smplrc
.... RECORD SPECIFICATION ....

            Organization   => 1
          1 Indexed   5 Redefined Record
          2 Relative   6 DL/I Segment
          3 Serial   7 SQL Row
          4 Working Storage

            Item Default Scope   => GLOBAL   GLOBAL, LOCAL
            File Name   => samplef
            Record ID Item   => keyfld
            Alternate Specification for =>
            File Location   =>

      For Variable Length Record:
            Record Length Item   =>
            Number of Occurrences Item =>

Figure 6.4.   Indexed record specification.

Record Specification—Indexed Record

Indexed records provide access to records by means of a key. Be sure that you have identified the position and length of the key field prior to beginning the definition process. See Figure 6.4. File name references a DD name in MVS, a DLBL in VM, or a CNAME in DPPX. This external name must be defined to the system being used before testing/execution will succeed.

Item Default Scope specifies default status of the record's items. Record ID Item names the key field. GLOBAL means that data items referenced may already exist in the current MSLs. If so, the definition can be used. If GLOBAL is specified and the data item is not in one of the current MSLs, CSP will add a definition to the current Read-Write MSL. LOCAL means that the data item definition is only stored as part of this record's definition. More information on GLOBAL and LOCAL is in-

cluded later in this chapter in the discussion of DATA ITEM
DEFINITIONs.

Alternate Specification For defines this Record ID as an alternate name to be used for the named Record ID (only Record
Specification and Prologue are allowed for Alternate names).
This allows you to use an existing record structure for different
files.

File location has meaning only in DPPX systems. It indicates
where a file may be found. The default value is blank. This indicates that the file exists on the local system.

For Variable Length Records, used to conserve disk space, the
length of record may be determined using one of the following
two methods:

Record Length Item

> a two-byte binary record length

Number of Occurrences Item

> numeric field in fixed-length portion of the record that contains
> the number of variably occurring segments contained in the
> record

Not all systems support variable-length records. Consult the
CSP documentation for your system to figure out whether or not
you may use them.

Record Specification—Relative Record

Relative records allow access by means of a relative record number. The file design usually dictates one of two methods to determine the key. Some systems use a data field with consecutive
increasing numbers as a key. Other systems use a "hashing" algorithm to calculate a record number based upon another value
stored in the record. You either have the perfect application for
relative files or you can't understand why in the world they exist!

See Figure 6.5. File name references a DD name in MVS, a
DLBL in VM, or a CNAME in DPPX. Item Default Scope specifies default status of the record's items. Record ID Item names a

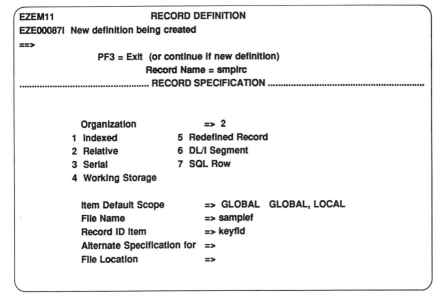

```
EZEM11                      RECORD DEFINITION
EZE000871  New definition being created
==>
                 PF3 = Exit  (or continue if new definition)
                        Record Name = smplrc
..................................... RECORD SPECIFICATION ..............................................


         Organization                 => 2
      1  Indexed               5  Redefined Record
      2  Relative              6  DL/I Segment
      3  Serial                7  SQL Row
      4  Working Storage

         Item Default Scope        => GLOBAL   GLOBAL, LOCAL
         File Name                 => samplef
         Record ID Item            => keyfld
         Alternate Specification for =>
         File Location             =>
```

Figure 6.5.   Relative record specification.

numeric data item (no decimals), used to specify the relative record number within the file. This data item is usually a Working Storage data item. Alternate Specification For allows specification of an alternate file name to be used as a model. File Location has meaning only in DPPX (see INDEXED record definition earlier).

Record Definition—Serial Record

Most execution environments use sequential files heavily. Sequential files may be defined to CSP as SERIAL records. CICS users should take care that serial records are defined properly to CICS if they are to be used.

See Figure 6.6. File Name references a DD name in MVS, a DLBL in VM, or a CNAME in DPPX. Item Default Scope specifies default status of the record's items. Alternate Specification for defines this Record ID as an alternate name to be used for the named Record ID (only Record Specification and Prologue are allowed for Alternate names since the data definition has al-

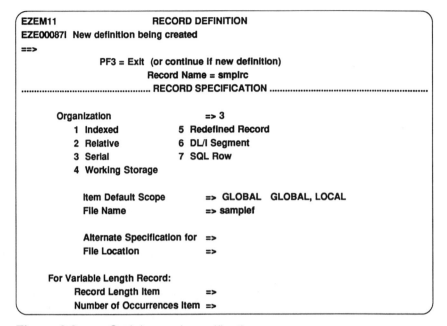

Figure 6.6.   Serial record specification.

ready been done). File Location has meaning only in DPPX (see indexed records). Variable-Length Records are sometimes used to conserve disk space. Length of record may be determined using one of two methods:

Record Length Item

a two-byte binary record length

Number Of Occurrences Item

numeric field in fixed-length portion of the record that contains the number of variably occurring segments contained in the record

Record Definition—Redefined Record

Occasionally, systems are designed using different record descriptions inside the same file. This practice was common in the days of cards (card type a, card type b) but it is unusual today. A

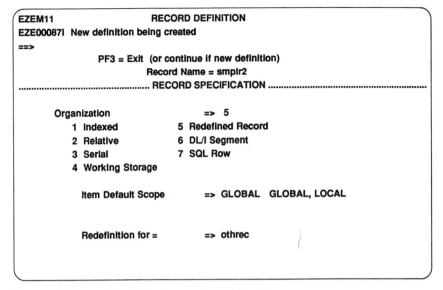

Figure 6.7.   Redefined record specification.

more likely use in a modern system is needing to have the same data known by several different field names. Should you need to define different records for the same file, use the REDEFINED RECORD organization type. REDEFINED records may not be referenced as objects in I-O processes. Data items within REDE-FINED record may be used by application processes if the record has been listed on the TABLE AND ADDITIONAL RECORDS LIST for the application.

See Figure 6.7. Name the record definition the redefined record is to represent. Remember that the redefined record may not be the *object* of a process but that redefined field names may be used if the additional record is defined to the application.

Record Definition—DL/I Segment

Once a PSB has been defined to CSP, the segments it references may be defined as CSP records. For each segment, a key must be defined that matches the database definition (Figure 6.8). Key Item names the segment key (again, it must match the definition in the database for position and length). Variable-Length Item is the

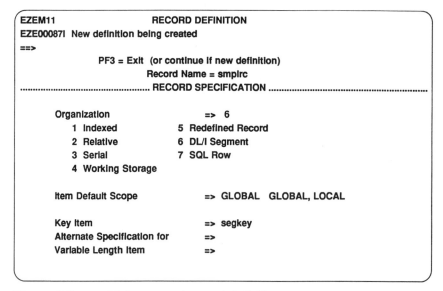

Figure 6.8.   DL/I segment record definition.

name of a field containing the segment length. Alternate Specification For names another record sharing the same structure.

Record Definition—SQL Row

Most installations using CSP are also using SQL heavily. Chapter 12 is devoted to the use of SQL. One of the more modern features of CSP is that for most I-O processes, it doesn't matter where the data actually resides. The CSP record definition takes care of that. For SQL Rows, CSP's record defines the *result* of a query. Savvy SQL users will adjust the default SQL created by CSP to exactly match the system's needs (Figure 6.9).

Alternate Specification For names another SQL row record using the same table(s) and column(s). SQL row records may only be alternate specifications for other SQL row records. List the table(s) to be used in the underlying query. SQL table names come in two parts, CREATOR and NAME, separated by a period (.). CREATOR is the ID used to create the table. If the current user created the table, then CREATOR may be omitted and SQL

```
┌─────────────────────────────────────────────────────────────────────────┐
│ EZEM11                    RECORD DEFINITION                              │
│ EZE000871 New definition being created                                  │
│ ==>                                                                     │
│            PF3 = Exit  (or continue if new definition)                  │
│                     Record Name = smplrc                                │
│ .............................. RECORD SPECIFICATION .....................│
│                                                                         │
│     Organization                  => 7                                  │
│          1 Indexed            5 Redefined Record                        │
│          2 Relative           6 DL/I Segment                            │
│          3 Serial             7 SQL Row                                  │
│          4 Working Storage                                              │
│                                                                         │
│     Item Default Scope            => GLOBAL   GLOBAL, LOCAL             │
│                                                                         │
│     Alternate Specification for   =>                                    │
│                                                                         │
│                                                                         │
│                                                                         │
│ .... Lines 00001 ...SQL Table Name(s)...................................│
│       TABLE NAME:                              TABLE LABEL:             │
│ ***                        TOP OF LIST                                  │
│ 001   dbuser.dbtab1                             T1                      │
│ ***                        END OF LIST                                  │
└─────────────────────────────────────────────────────────────────────────┘
```

Figure 6.9.   SQL row record definition.

will assume it. Table NAMEs must be keyed in regardless. Some installations will use SYNONYM names for tables. This allows an individual to use a table without having to key in (or even know!) the actual CREATOR ID or table NAME. In the list of tables, CREATOR_ID.TABLE_NAMEs or SYNONYM NAMEs are used to create the SQL row. TABLE LABEL is an abbreviated label used to refer to CREATOR.TABLE (SQL alias/correlation name) in the SQL statement(s) constructed by CSP. It is a good idea to modify the TABLE LABEL to something meaningful like the table NAME. As many tables as are required for the underlying query may be named. "Wait!" I hear all you SQL users saying. "What about the JOIN condition?" It is your responsibility to check the default SQL statement generated using the DEFAULT SELECTION CONDITIONS DEFINITION panel. This will be explained fully in Chapter 12.

Record Definition—Data Item Definition

Once the record has been defined, the data items (fields) that are part of the record may be defined. Be especially careful to pay attention to the GLOBAL or LOCAL aspect of data item definition (Figure 6.10). Failure to plan can cause serious problems later.

```
EZEM12                        RECORD DEFINITION                    More ->

==>

                              PF3 = Exit
                              Record Name = smplrc          Default Scope = LOCAL
Total lines 0008...........................DATA ITEM DEFINITION .........................................
    NAME                      LEVEL OCCURS TYPE LENGTH DEC      BYTES  SCOPE
    ***                       TOP OF LIST
    001 smprec                10    00001   CHA  00150          00150  GLOBAL
    002 keyfld                15    00001   CHA  00015          00015  GLOBAL
    003 field1                15    00001   PACK 00005    0002  00003  LOCAL
    004 fdate                 15    00001   CHA  00006          00006  GLOBAL
    005 fmm                   20    00001   CHA  00002          00002  GLOBAL
    006 fdd                   20    00001   CHA  00002          00002  GLOBAL
    007 fyy                   20    00001   CHA  00002          00002  GLOBAL
    008 *                     15    00001   CHA  00126          00126  LOCAL
```

Figure 6.10. Data item definition.

The DATA ITEM DEFINITION panel is used despite the data's organization type. Please notice the "More —>" arrow in the upper right-hand corner of the panel. At various places within CSP/AD, when multiple screens are available, the arrows tell you which way(s) you may scroll. Use the LEFT (PF10) or RIGHT (PF11) to scroll in the desired direction. Shifting right allows entry of descriptions (Documentation!—What's he trying to pull?) for the data items. Each column on the DATA ITEM DEFINITION panel is examined in detail over the following paragraphs. Please refer to the CSP documentation if you have further questions.

NAME is the field name to be used by CSP. Field names may be 1 to 32 characters in length. The first character must be A–Z, $, #, or @. The first three characters *may not* be EZE. Other characters may be A–Z, 0–9, $, #, @, –, or _. A single asterisk (*) may be used instead of a field name to save unnamed space in the record (COBOL programmers call this FILLER). Before Version 3 Release 3 of CSP names were limited to eight characters and did not allow "hyphens" (–) or underscores (_). Be careful! Special characters are converted to hyphens when COBOL source code is generated by CSP. If your application will be using COBOL source generation (if you aren't now, you will be when Version 4 comes along), you should probably avoid using special characters other than hyphens. An exception to this might be SQL data items. SQL column names may use underscores but not hyphens. It sometimes works to set standards so that CSP field names are identical to SQL column names *except* that the underscores are changed to hyphens.

LEVEL refers to the relative nesting of the data item within the record definition. Lower numbers represent group items that include all higher-level items until the next lower-level number. CSP allows you to specify levels 03 through 49. Levels 01, 02, and those higher than 49 (except 77) are not allowed. CSP uses the first two levels internally. Levels are frequently used to identify structures and substructures in the data. In the example in Figure 6.10, the entire record is described by the LEVEL 10 data item named "smprec." The same example shows fields "fmm," "fdd," and "fyy" as part of a structure called "fdate." You may use LEVEL 77 for data items, but limitations regarding LEVEL 77 item use when transferring control may prove bothersome (see Chapter 13).

OCCURS is a number representing the number of repetitions for a field. CSP allows 1 to 32767 occurrences. Only one-dimensional tables are allowed by CSP. Subscripting and use of arrays is covered in Chapter 10.

TYPE describes the data format. If is important whenever data will be passed to other programs or when COBOL is generated that the data types be compatible. The types of data supported are listed below:

Type	Description
BIN	Binary data
CHA	Character data
DBCS	Double Byte Character Set data
HEX	Hexadecimal data
MIX	Combined DBCS and CHAracter data
NUM	Numeric characters, all positive, signed with x'F' (zoned/external decimal, COBOL DISPLAY)
NUMC	Numeric characters, positive signed with x'C', negative signed with x'D' (zoned/external decimal, COBOL DISPLAY)
PACK	Packed-decimal data signed with x'C' for positive values and with x'D' for negative values
PACF	Packed-decimal data, all positive, signed with x'F'

LENGTH describes the maximum number of characters or digits to be held by the field. LENGTH does not necessarily describe the storage required. That is why the BYTES column exists. The maximum value in the LENGTH column varies based upon data type.

Type	Length
BIN	1–18 (9 for OS/400)
	1–4 digits represents a half-word in MVS (2 bytes)
	5–9 digits represents a full word in MVS (4 bytes)
	10–18 represents a double word in MVS (8 bytes)
CHA	1–32767 (254 for items in tables)
DBCS	1–16383 (127 for items in tables)
HEX	1–65434 (254 for items in tables)
MIX	1–32767 (254 for items in tables)

NUM 1–18 (same for NUMC)

PACK 1–18 (same for PACF)

size in bytes = (#digits/2) + 1 (no rounding)

Specify the length of a group (structure) data item as zero and press PF3. CSP will calculate the size of the group for you. CSP always verifies the length of group items before saving when the PF3 key is pressed. A good practice is to define a single 05 LEVEL data item that describes the entire record. This helps make sure that your definition is error-free.

The DEC column shows the number of decimal places in a number. Decimals are included in the complete field length (though the actual decimal sign is implied). For instance, a field necessary to hold amounts of U.S. currency up to a maximum of $5 would have LENGTH=3 and DEC=2. The number of decimal places may not exceed field length.

BYTES shows the actual space required to store the field. CSP will calculate the number of bytes based upon data type and length. If you modify BYTES for a known column, CSP will modify the field length based upon the data type to match.

Beginning with Version 3 Release 3 a new column called SCOPE is present when defining data items. SCOPE allows the developer to decide whether to share common data definitions or not. By marking SCOPE appropriately, a developer may make the occurrence of a record's data item LOCAL (rather than GLO-BAL) to the record definition and not shared across the MSL. LOCAL data items will help reduce inadvertent "collisions" due to nonexistent, misunderstood, or violated naming conventions. Map fields are always LOCAL (though they may initially create or use a GLOBAL definition). Record and table data items may be LOCAL or GLOBAL. GLOBAL characteristics are stored in an MSL DATA ITEM member. LOCAL characteristics are stored as part of the record's or table's MSL member.

The question to use GLOBAL versus LOCAL data items is an important design decision (all data items were GLOBAL for CSP releases before Version 3 Release 3 (except map data items)). GLOBAL data items are stored individually in the MSL; their definition and properties are shared.

When you use a data item name for the first time while defining a record, CSP/AD searches the MSL concatenation for a data item using the same name. If one is found, CSP/AD asks the user to press ENTER to accept the existing GLOBAL definition. The new record simply POINTS TO the GLOBAL definition. This allows the most currently available definition to be used. *Warning:* If a GLOBAL field is referenced when none of the current MSLs contain a definition for the data item, CSP/AD uses a default data definition for the field (CHA 3).

LOCAL data items are known only to the object in which they are defined. They must be individually defined for each object (though you may start by calling a field GLOBAL to copy a definition, then modify it to LOCAL later).

GLOBAL definitions allow use of naming conventions and a data dictionary approach to planning a system. LOCAL data items guard against the accidental deletion of a GLOBAL definition or MSL—at the cost of redundant (perhaps incorrect) data definition and increased maintenance costs when data definitions must change. Is there a "best" way? It depends on how solid the standards used at your shop are. If a strong set of programming standards and data dictionary are present, then without question a GLOBAL approach is the way to go. If your organization is like many others, standards are a target rather than a reality. Without firm standards, GLOBAL variables can sometimes lead to trouble. A well-thought-out plan should be followed (again, there is no substitute for good design).

Record Definition—Record Prologue

CSP provides a built-in mechanism for documenting a record definition. Creation of a prologue is not part of the automatic definition cycle for CSP records. You must request Option 3 from the RECORD DEFINITION FUNCTION SELECTION panel (Figure 6.11). The prologue is a separate part of documentation in addition to the descriptions allowed for each field. The prologue will be displayed any time someone requests it, and it will be printed any time the record definition is printed.

Each line in the prologue may be up to 60 characters long.

```
EZEM04                    PROLOGUE DEFINITION

==>
                             PF3 = Exit
                           Member = smplrc
Total lines 0003 ...................... MEMBER PROLOGUE ...............................................................
***                    TOP OF LIST
001 Here lies a comment...                                              *
002 This is yet another line...                                        *
-->                                                                     *
***                    END OF LIST
```

Figure 6.11.   Record definition prologue.

There is no size limit for prologues, so you may enter as many lines as necessary. Whenever deciding how to document something, ask yourself this question, "What would I be most comfortable with if two years from now I have to fix this on the afternoon of the day before I go on vacation?" A record definition's documentation should include at least the name of the creator, date of definition, group responsible for maintenance, a description of record's purpose, and the name of the host system files the record represents. As fields are added or altered they should be documented in the prologue.

Record Definition—Default Selection Conditions Definition

When you define an SQL row, the default query built by CSP is shown on this panel. This part of an SQL row is not part of the automatic definition cycle for CSP records. You must request Option 4 from the RECORD DEFINITION FUNCTION SELECTION panel (Figure 6.12) to get here. This panel has meaning only for SQL rows.

On this panel you may alter the default statement's WHERE clause. It is essential that you make sure that JOIN conditions are appropriately identified for any statements using more than one table. If you do not have SQL experience, ask someone for

```
EZEM16                    SQL ROW RECORD DEFINITION
EZ005901 You may edit lines preceded by line numbers
==>
  PF3 = File and exit  PF4 = Reset to default statement   PF5 = SQL syntax check
  Record = smplrow
                                          Modified clause      = NO
Total lines 0009 .......... DEFAULT SELECTION CONDITIONS DEFINITION ................................
                          TOP OF LIST
***  SELECT
***     COL1, COL2, COL3
***  INTO
***     :COL1, :COL2, :COL3
***  FROM
***     CREATOR.TABLEA T1,
***     CREATOR.TABLEB T2
***  WHERE
009 ;** INSERT DEFAULT SELECT CONDITIONS HERE **
***                      END OF LIST
```

Figure 6.12.   Record definition—SQL row record default selection conditions.

help. It is important to remember that this is the default SQL statement for the SQL row. When processes are defined using the SQL row record as an object, CSP will build a new SQL statement for that process using this as its basis. The developer may alter the individual SQL statement after the process is defined (see Chapter 12 for specifics about SQL use).

FILE ALLOCATIONS

Before files may actually be used, they must be available to the environment that CSP is operating in. This means that the FILE field in the record definition must be a name known to the host environment. For instance, under TSO ALLOCATE statements must have been issued, under VM DLBL statements are required, and CICS requires entries in the FCT. If you are unfamiliar with the terms used above, find a technical person to help you.

CHAPTER 6 EXERCISES

Questions

1. What command may be keyed on any command line to go immediately to the main definition panel?
2. How is a CSP record definition related to files in the operating system for testing and execution?
3. Do all CSP record definitions require keys?
4. Name the CSP record type most commonly associated with VSAM Key Sequenced Data Sets (KSDS).
5. How long may a CSP data item's name be?
6. What is entered as a field name when you don't want to name the field but you want the definition to reserve space?
7. When defining an SQL row, what is entered as in the TABLE NAME column for a table named 'mothra' belonging to a user named 'godzilla'?
8. What is the maximum number of lines in a record prologue?
9. What would the LENGTH and DECIMAL values be for a field that could hold the value 123456.789?
10. What happens if you set a group data item's length to zero?

Computer Exercise
Define a VSAM Record

Define the following record to CSP. It is used in coming exercises.

1. Record name: BB01R01   File name (DDNAME): BOOKS
2. This is an INDEXED record, representing a VSAM KSDS
3. The record description follows (fixed length, 220 bytes long):

Data Field	Positions	Bytes	Definition
INVENTORY-ID	1 - 12	12	CHA
ISBN	13–22	10	CHA
PUBLISHER	23–62	40	CHA
AUTHOR	63–102	40	CHA
TITLE	103–142	40	CHA

Data Field	Positions	Bytes	Definition
unused	143–143	1	CHA
REPLACEMENT-COST	144–146	3	PACK 999V99
NORMAL-PRICE	147–149	3	PACK 999V99
SALE-PRICE	150–152	3	PACK 999V99
unused	153–156	4	CHA
QUANTITY-ON-HAND	157–158	2	BIN 9999
QUANTITY-ON-ORDER	159–160	2	BIN 9999
unused	161–161	1	CHA
DATE-FIRST-EDITION	162–167	6	CHA (yymmdd)
DATE-CURR-PRINTING	168–173	6	CHA (yymmdd)
BOOKCOVER	174–174	1	CHA
DEWEY-NUMBER	175–177	3	PACK 999V99
LIB-CONGRESS-CAT	178–197	20	CHA
LIB-CONGRESS-NBR	198–204	7	CHA
unused	204–220	16	CHA

7

List Processor

By now, you may have been looking for a quicker way to get to CSP objects you wish to change. As if in anticipation of your desires, the designers of CSP have provided the List Processor. Heavy users of CSP/AD often base their activities from the List Processor panel. The List Processor can list all or part of the contents of one or more MSLs.

To reach the List Processor, simply enter option 1 from CSP's FACILITY SELECTION "main menu" (Figure 7.1), or go directly into the List Processor by entering "=List" from the command line on any panel. The List Processor is dependent upon the current MSL concatenation for data. If a name appears in more than one eligible MSL, only the *first* MSL's object is shown. Without interrogating the other MSLs, there is no way to know if an object resides in more than one MSL. The List Processor has many settings that may be set to control the scope of the list. The broader the scope, the longer it will take CSP to create the list. If you are sharing large MSLs, it is a good idea to narrow the search carefully to save time. If the MSL is not tuned well, the search will be slower than necessary. It is a good idea to reorganize the MSL files frequently.

```
EZEM00          CROSS SYSTEM PRODUCT/APPLICATION DEVELOPMENT

==>
                        ENTER = Continue  PF3 = Exit
R/W MSL => yourmsl                                       HIGHEST MSL# =>1
.............................................. FACILITY SELECTION ..............................................


                Enter number of facility desired => 1

                    1    List Processor
                    2    Definition
                    3    Test
                    4    Generation
                    5    Utilities and File Maintenance
                    6    Tutorial
                    7    MSL Selection

                In all facilities:
                PF1 = Help  PA2 = Cancel
```

Figure 7.1.   Facility selection.

The List Processor (Figure 7.2) uses search settings for object
NAME, object TYPE, and MSL number to limit the list. CSP/AD
has built-in default values for the List Processor search settings. By
simply pressing ENTER (accepting the defaults), you list all object
names from the first MSL in the concatenation except data items.

```
EXEM76                       LIST PROCESSOR

==>
            ENTER = Continue      PF3 = Exit      PF4 = Refresh
        SUBSET:  Name =>    *      *
                 Type =>   -I        MSL # =>   1    Limit =>    0200
    Options:  Edit  = E    Copy  = C   Rename    = R   Where used= W  Change= 1
              View  = V    Omit  = O   Current   = /   Associates = A  Export= X
              Delete = D   Test  = T   Print     = P   Generate   = G
Total lines 0000 ...................................... MEMBER LIST ............................. Lines 0001 to 0001
OPT  NAME                            TYPE MSL      MODIFICATION / OPERANDS
                                  TOP OF LIST
                                  END OF LIST
```

Figure 7.2.   List processor.

To look at another MSL, change the MSL# value to match the desired MSL's relative position in the current concatenation sequence (MSL list is option 7 from main menu or =MS). Later in this chapter we'll cover the use of the MSL SELECTION panel for this. To let the List Processor look at all MSLs in the current concatenation sequence, change the MSL# to a single asterisk (*) before beginning the search (the asterisk may be used as a "wild card" for all List Processor options).

Be careful! If the List Processor is looking at all currently available MSLs, only the first reference found to any particular object name is listed. Be sure to check the MSL number displayed in the list to see which MSL an object came from. CSP will *not* tell you if the same object occurs in other MSLs further down the list.

To search for specific object names, modify the "NAME=>" setting. Since Maps within Map Groups are known by both names, both may be entered (group first, then map, separated by a blank). For instance, to find the map created in the first exercise, enter "bb01g bb01m01" in the name setting and press ENTER. The asterisk (wild card) may also be used in names. To find all objects beginning with "bb," enter "bb*" and press ENTER. Notice that the default value is a single asterisk, meaning that *all* names are to be listed. Here are some other examples:

TZ* displays all MSL items that begin with TZ.
*GP displays all MSL items that end with GP.
X10 displays all MSL items with the "X10" in their name.

TYPE is used to control the type of object listed. The list may return a single object type, all object types, or object types not specifically omitted from the list.

To choose only objects from a specific type, enter the correct value:

Type	Object
APPL	Applications
ITEM	Items
MAP	Maps

MAPG   Map Groups (when floating maps are used)

PROC   Processes

PSB   DL/I PSBs (Program Specification Blocks)

RECD   Records

SGRP   Statement groups

TBLE   Tables

To list all object types, enter a single asterisk for type. To omit selected types from the list, key in a hyphen (minus sign, "–") followed by the first character of the definition type(s). For example: "TYPE => –APS" removes Applications, Processes, PSBs, and Statement Groups from the list.

LIMIT is the maximum number of things to list in your session. Using LIMIT may greatly reduce the time consumed by search settings that are too broad.

Once you have keyed in the search settings and pressed ENTER, a list something like Figure 7.3 appears.

CSP provides OPTions (listed at top of screen) that allow different actions with items in the list. The OPT column allows you

```
 EXEM76                        LIST PROCESSOR

 ==>
                 ENTER = Continue      PF3 = Exit       PF4 = Refresh
          SUBSET:  Name =>   bb01**
                   Type =>   -I          MSL # =>   1     Limit =>   0200
     Options:   Edit   = E    Copy  = C   Rename   = R   Where used= W   Change= 1
                View   = V    Omit  = O   Current  = /   Associates = A  Export= X
                Delete = D    Test  = T   Print    = P   Generate   = G
 Total lines 0000 ........................ MEMBER LIST ............................ Lines 0001 to 0001
 OPT  NAME                          TYPE MSL        MODIFICATION / OPERANDS
                                TOP OF LIST
      BB01A                         APPL  1   10/31/91    13:13
      BB01G     BB01M01             MAP   1   10/31/91    12:01
      BB01R01                       RECD  1   10/31/91    14:03
                                END OF LIST
```

Figure 7.3.   List processor (with data).

to specify what you want to do with the objects listed (see chart below).

The MODIFICATION/OPERANDS column provides a place to enter new names when using the Copy, Export, or Rename facilities. Be sure to use the Erase EOF or space bar to clear the rest of the field before pressing ENTER.

Options provided by the List Processor and their meanings are listed below. For more complete information refer to the CSP/AD Developing Applications manual.

OPT	Meaning
Edit	Modify chosen object via definition panels (automatically shifts to view mode for read-only MSLs)
View	Display object definition panels (except prologue)
Copy	Copy object, must list new name under OPERANDS (for maps, both map group and map name are required), items copied from read-only MSLs are copied without change;
	to replace a member in the read-write MSL with a like named member from a read-only MSL, specify REP under OPERANDS
	Warning: When copying record definitions, only LOCAL data items are copied; pointers to the current MSL concatenation are used for GLOBAL data items. Likewise, processes are NOT copied when applications are.
Omit	Hide chosen objects from current list (no permanent impact)
Test	Begin testing of selected application
Delete	Delete chosen object
Print	Print object plus associates (printer must be defined to CSP)

Rename	Change name of object; must list new name under OPERANDS (for maps, both map group and map name are required)
Current (/)	Places object at top of current displays
Change (1)	Renames object to the name specified under OPERANDS everywhere it occurs in the current read-write MSL except comments, prologues, and descriptions (map group names are not changed either, only map names). This option searches through *all* objects in the current MSL and is less efficient than using the CHGList command (see below)
Export	Invokes MSL utility to export the named object (without associates)
Where	Generates a new list of MSL members containing references used to the specified object
	Caution: Where looks at *all* concatenated MSLs and may require a great deal of time
Associates	Lists all associates of this member that would be exported by an EXPORT
Generate	Invokes the MSL generate utility for the named application, map, map group, or table

The CHGList command may be used instead of the Change (1) option. CHGList may be considerably faster since it searches only the objects shown currently by the List Processor. Limit the List Processor's object list searched by using the wild card names, WHERE USED (W), ASSOCIATES (A), and HIDE (H) options.

The old object name (oldobjnam in Figure 7.4) is replaced by the new object name (newobjnam in Figure 7.4) throughout the objects in the current list. Read-only MSLs and objects not in the current MSL list are not affected. The item type is specified to further limit the processing required. If options (OPT) were entered at the same time as CHGLIST, the options are executed

---

CHGList oldobjnam newobjnam

                                APPL or ITEM or
                                MAP or PROC or
                                PSB or RECD or
                                SGRP or TBLE

- CSP searches for "oldobjnam" in Read-Write MSL objects in the current List Processor list and changes it to "newobjnam" everywhere it occurs

- Object type is required if the oldobjnam being searched for references a LOCAL data item and a non-item object with the same name exists already in the MSL

Figure 7.4.   CHGLIST command syntax.

first, followed by CHGLIST. CSP list displays "*** Changed ***" next to object altered by CHGLIST.

WHERE USED AND ASSOCIATES

The LIST facility provides the capability to list all other MSL items that use a specified item (W = Where used). Each MSL item that directly references the specified item is listed. By using the W (WHERE) option for each successively higher-level item, a complete list of all items impacted by the original item can be created.

The LIST facility also provides the capability to list just the MSL items that are used as a part of a given item (A = Associates). Each MSL item and its subordinate items are listed. The list of items that shows as ASSOCIATES for an application reflect the MSL items that will be included in an EXPORT or GENERATE of the application. There is, of course, one exception: In more than one system checked by this author ASSOCIATES didn't list a Statement Group that was used only in an automatic map edit. I think this is a "bug" since it is undocumented. (For those who are unindoctrinated, if it's documented, it's a *feature*, not a bug!) Careful use of a naming convention allows use of

CSP's LIST facility as a data dictionary-type aid. If a naming convention is being followed, WHERE USED and ASSOCIATES are particularly useful.

LIST PROCESSOR AND MSL CONCATENATION

Items displayed by the LIST processor reflect the FIRST item of a particular name found in the concatenation of MSLs known to the CSP/AD session. The MSL sequence specified on the MSL selection panel (Figure 7.5) controls which MSLs items are listed from. ONLY the first MSL item found in the concatenation of libraries is listed (unless MSL# specified in list or MSL#=>* is used).

Be careful! Because each item named in a CSP application must either be defined as LOCAL to a defined entity or default to GLOBAL, it is possible to accidentally inherit/change the attributes of an MSL item that is used elsewhere.

```
 EZEM01                        CROSS SYSTEM PRODUCT

 ==>
                     ENTER = Continue      PF3 = File and Exit

 ................................ MSL SELECTION ................................

              MSL FILE NAME      MSL ACCESS        CONNECT
                                   ORDER           STATUS

 Read/Write MSL:
        => yourmsl               => 1

 Read-Only MSLs:
        => romslxx               => 2
        => romslaa               => 3
        =>                       => 4
        =>                       => 5
        =>                       => 6


                          In all facilities:
                      PF1 = Help     PA2 = Cancel
```

Figure 7.5.   MSL concatenation.

To get to the MSL SELECTION list, either use option 7 from the FACILITY SELECTION "main menu" panel, or enter the "=MSl" command from any command line. MSLs may be added by simply keying in their names on this panel. Modify the MSL ACCESS ORDER numbers to alter concatenation sequence (or you could rekey all the file names instead if you like hard work). Blank out any MSLs you wish to omit (the Read/Write MSL is optional for reading and testing). Don't forget that any files referenced must have already been ALLOCATEd for the system CSP/ AD is running in. For instance, CICS users must make sure that an FCT entry exists for the defined name, TSO users must make sure an ALLOCATE statement has been executed, and VM sessions must have defined the DLBL.

CSP/AD FACILITY TRANSFER COMMANDS

Anticipating the need to back in and out of CSP/AD functions when you want to check a definition or add a field, CSP has provided a shortcut feature between functions! CSP provides several commands that allow you to take "shortcuts" through CSP/AD called Facility Transfer commands. Facility Transfer commands were new with CSP/AD Version 3 Release 3. Each of the commands is prefaced by an equal "=" sign. Those familiar with ISPF/ PDF will recognize the idea, but this is a little different.

Keyed into the command line as a primary command, the facility transfer commands cause an immediate branch to the named facility. The facility to transfer to is named, and the transfer name is preceded by an equal sign (=, a la ISPF). Figure 7.6 shows how to branch directly to the List Processor.

=List

Figure 7.6.   Sample facility transfer statement.

The syntax for facility transfer commands is listed in Figure 7.7. The commands each cause transfer to a different facility. One drawback: CSP does not remember the name of the object you were using before the transfer operation. By default, the facility transfer commands all use the option FILE to cause

```
=COpy
=Delete
=DIrectory
=Edit
=EXport
=FChange
=Fview
=Generate
=Import
=List
=Main
=MSI
=Print
=Rename
=Test
      File or Cancel
```

- **Commands each cause transfer to different CSP facility**

- **File or Cancel after transfer command to SAVE or CANCEL current panel's data (File is the default)**

Figure 7.7.   Facility transfer syntax.

whatever you are doing on the current screen to be saved before the branch. To branch without saving, key in the CANCEL option (like pressing PA2 before branching). If there is an error on the current panel, the transfer is canceled and must be rekeyed after the error is fixed.

The different Facility Transfer options are listed below:

Facility Transfer	Where It Takes You
=COpy	Copy MSL member
=Delete	Delete MSL member
=DIrectory	Print MSL directory
=Edit	Go to definition

=EXport	Export MSL member
=FChange	Display file records for update
=Fview	Display file records
=Generate	Generate the MSL member
=Import	Import an MSL member
=List	Branch to List Processor
=Main	Go to main menu, facility selection
=MSI	MSL selection panel
=Print	Print member
=Rename	Rename member
=Test	Begin test facility

CHAPTER 7 EXERCISES

Questions

1. Name both ways to get to the List Processor.
2. From the List Processor, what option is used to edit an object?
3. What List Processor option allows you to see the names of objects that *directly* reference an object?
4. Which List Processor option prints an object?
5. What List Processor option shows the names of all objects to be included when an object is printed or generated?
6. Which facility transfer command is used to return to the FACILITY SELECTION "main menu" panel?
7. Name the facility transfer command used to go to the MSL list.
8. Will a transfer work if there are errors on the panel being transferred from?
9. What command is issued to go directly to the List Processor *without* saving the most recent changes on the current panel?

10. What command is issued to save the most recent changes on the current panel and then go to DEFINITION SELECTION panel?

Computer Exercise
Using the List Facility and Transfer Commands

1. Use the List Processor to show the contents of your Read-Write MSL.
2. Modify the MSL concatenation so that there is no Read-Write MSL, an MSL called "UTILMSL" is MSL number 2, and your MSL is MSL number 3.
3. List all records from both MSLs.
4. Modify the MSL concatenation so that your MSL is once again the Read-Write MSL.
5. List all data items in your MSL.
6. List all objects using the data item INVENTORY-ID (or whatever you called it in your definitions).
7. Print your application.
(Successful printing may require involving your CSP administrator to make sure a printer is defined.)

8

Direct File Access

Well, you must be ready to do more with CSP! This chapter will walk you through the steps necessary to do basic I-O operations. You will learn how to read records (INQUIRY), read records for the purpose of updating them (UPDATE), modify records (RE-PLACE), insert new records (ADD), and delete records (DE-LETE).

But first, a short review of some basic ideas and terms we've covered so far.

Application	CSP "program," collection of Processes and statement groups used to solve some business need
Processes	Blocks of CSP code, usually associated with some I-O process
Statement groups	Optional blocks of CSP code
Statements	Found inside of Processes and statement groups, provide data movement and manipulation
Flow statements	"Go to" logic used to control processing of older systems; to be avoided in new systems

Structure list	Result of pressing PF4 from the APPLICATION PROCESS LIST, or of choosing option 5 from the APPLICATION DEFINITION FUNCTION SELECTION panel
Object definition	Definition of RECORDS, MAPS, and DATA ITEMS to be used by Processes
Error processes	Statement groups to be executed in the event of I-O error; the special CSP word EZERTN is often used to cause return to the Process's logic.
Record definition	Definition of records to be used by I-O processes
Map definition	Definition of maps to be used by I-O processes

Application processing always starts with the first process in the APPLICATION PROCESS LIST and "falls through" all other processes at the highest level until done. Branching is possible using the FLOW sections of the highest-level processes, but it is not a good idea. Most installations advocate the use of a single highest-level module that executes subordinate Processes and statement groups in a branch and return fashion. The branch and return technique seems more consistent with modern "structured" design concepts and provides more easily maintained code.

Processes represent a basic unit of CSP code. Most processes are associated with record or map I-O. Processes are SHARED by all who have access to the MSL they are defined in and who add that MSL to their concatenation list.

Processes that involve I-O are split into two halves. The first half of an I-O process includes code that is executed *before* the I-O (like setting key values). The second half of an I-O process has code that executes *after* the I-O (perhaps doing something with the data). EXECUTE processes do not perform I-O, so they are not split. Processes entered on the APPLICATION PROCESS LIST are at level 1 in the application structure. These processes are called "main" processes. Most installation standards call for a single main process, the so-called "mainline" routine.

Many processes (those involving records) require the naming

of error routines or EZE operations to be executed if an I-O raises a non-zero return code. If the error routine is a statement group, CSP executes it in a branch and return fashion with execution resuming after the I-O in the process. It is common to use EZERTN in the place of the error routine. EZERTN does nothing and returns control to the beginning of the Process's *after* I-O logic. If the process error routine names a main process, a direct branch (go to) is executed upon error (this is not a recommended approach). Finally, some other EZE word, like EZECLOS, might be used to cause the immediate end of an application in the event of an error. Failure to code an error routine name will cause immediate end of the application if an I-O raises a non-zero return code (indicating some kind of warning or error).

Each Process includes a Flow section. If used, the Flow section controls the flow of the application by means of direct branching. If multiple main processes (processes at level 1) are defined, execution will fall through from one to the next in sequence. This fall-through is controlled by using the Flow section's direct branch capabilities. Most shops have outlawed using the Flow section for new development in favor of PERFORMing processes. Figure 8.1 illustrates the flow of CSP process logic.

CSP PROCESS LOGIC FLOWS IN THIS ORDER:

- 1 Code executed BEFORE process option

- 2 PROCESS OPTION

- 3 If error detected, error routine is executed

- 4 Code executed AFTER process option

- 5 FLOW section

    (FLOW branching commands only work for main processes and are forbidden in most shops anyway)

Figure 8.1.   CSP process logic sequence.

APPLICATION DESIGN AND DEFINITION

By now, the need to document CSP applications and design beforehand is probably becoming obvious. Don't forget to use the PROLOGUE to document your intentions, decisions, and changes. Make sure that the design is modular and detailed at least to the point where individual I/O activities are identified.

Each module in your design will represent (1) I/O activity using various PROCESS OPTIONs, or (2) control activity using either EXECUTE PROCESS OPTIONs or statement groups. Be sure to document the structure! Figure 8.2 illustrates a common device called a "Structure Chart." Each box in the structure chart will be represented by a Process or statement group in the CSP application. Sometimes modules are used repeatedly. The box marked "I-O PROCESS" shows one mechanism for illustrating that a module is used more than once.

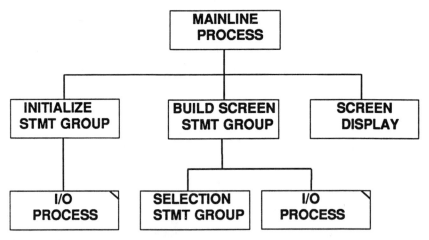

Figure 8.2.   Structure chart.

It is important to realize that structure charts are just a design tool. If you are more comfortable using some kind of brackets, braces, bubbles, or some other technique to document your design, then don't change now! The point is that the design should be thought out and documented.

The steps to defining an application to CSP are the following:

1. Design the application.
2. Define the application to CSP using the available screens (be

careful not to add unnecessary processes). The mainline logic and all I/O routines must be defined using CSP PROCESSes. Logic will either be included inside the logic of mainline EXECUTE PROCESSes, as part of the I/O PROCESS OPTIONs and their logic, in separate EXECUTE PROCESS OPTIONs, or statement groups.

3. Decide which process options to use to match your design.
4. For each PROCESS OPTION decide on a name for the process, which object it will use, and what error routine should be named (if any).
5. Code the processes and their statements.

Don't forget that I-O processes are divided by the I-O activity. Figure 8.3 shows the execution sequence of an I-O process graphically.

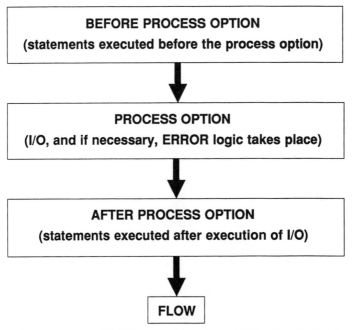

Note: In most shops FLOW processing is prohibited or limited to a single flow statement at the end of the main routine which FLOWs back to the top of the main routine. This eliminates the need to DO WHILE for the main routine's looping (FLOW is only meaningful for top-level processes).

Figure 8-3. Process option execution sequence.

APPLICATION DEFINITION

Application definition itself was discussed in Chapter 4. Refer to Chapter 4 to refresh your memory on getting into Application Definition, or place an "E" next to an application name on the List Processor and press ENTER.

The basic terms of process definition can stand review at this point as well. Using the APPLICATION PROCESS LIST, the mainline process was named (Figure 8.4). Place a "P" in the SEL column next to the mainline routine to see the code associated with it. This routine should either PERFORM I-O routines or routines that do so. Return to the APPLICATION PROCESS LIST by pressing PF3. By pressing PF4 from the APPLICATION PROCESS LIST or by initially selecting STRUCTURE LIST, you cause all of the application's processes and statement groups to be listed. In the column marked LVL, CSP indicates the relative placement of each routine in the application's logic. The "mainline" routine should be the only one at level 1. Lower-level routines are listed directly below the higher-level routine that causes them to be executed.

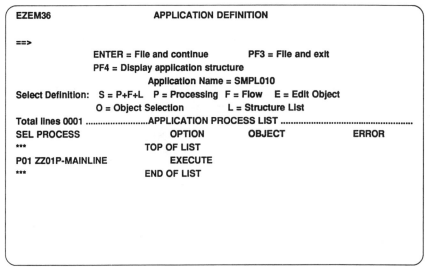

Figure 8.4. Application process list.

```
┌─────────────────────────────────────────────────────────────────────┐
│  EZEM36                    APPLICATION DEFINITION                     │
│                                                                       │
│  ==>                                                                  │
│              ENTER = File and continue    PF3 = File and exit    PF4 = Refresh │
│                            Member Name = ZZ01A                        │
│  Select Definition:   S = P+F    P = Processing    F = Flow    E = Edit Object │
│                    O = Object Selection                               │
│  Total lines 0001 ......................PROCESS AND GROUP LIST ....................................... │
│  SEL PROCESS          LVL  OPTION        OBJECT            ERROR      │
│  ***          TOP OF LIST                                             │
│  001 ZZ01P-MAINLINE   001  EXECUTE                                    │
│  001 ZZ01P-CONVMAP    002  CONVERSE      ZZ01M01                      │
│  001 ZZ01P-EDIT       002  EXECUTE                                    │
│  001 ZZ01P-GETREC     003  INQUIRY       ZZ01R01           EZERTN     │
│  001 ZZ01P-WHOOPS     004  EXECUTE                                    │
│  001 ZZ01P-CALCTAX    004  EXECUTE                                    │
│  ***                  END OF LIST                                     │
│                                                                       │
└─────────────────────────────────────────────────────────────────────┘
```

Figure 8.5.   Structure list.

In the structure list illustrated in Figure 8.5, the mainline routine ZZ01P-MAINLINE causes two routines to be executed, ZZ01P-CONVMAP, and ZZ01P-EDIT. The ZZ01P-EDIT routine causes the ZZ01P-GETREC routine to be executed. ZZ01P-GETREC executes ZZ01P-WHOOPS or ZZ01P-CALCTAX. The levels illustrated on the structure list should map directly to your design documentation when you are finished. Placing a "P" in the SEL column next to the ZZ01P-GETREC routine causes its logic to be displayed. Notice that this process uses the process option INQUIRY to read data from the file described to CSP by the record definition ZZ01R01(Figure 8.6).

This routine moves data into the record's key field (conveniently named, don't you think?) in the area *before* the process option does the I-O. If an error happens, EZERTN is executed, returning control to the *after* portion of the logic. The process then checks the outcome of the I-O. NRF (No Record Found) means that the key was not found on the file; ERR means any non-zero response. Please notice the documentation shows the developer's intent. See Chapter 5 or the CSP documentation for more information about error condition tests.

```
EZEM39                 APPLICATION PROCESS DEFINITION

==>
                PF3 = File and exit  (or file and continue if more selected)
        Process => ZZ01P-GETREC          Description = Get a record
        Option = INQUIRY                 Object  = ZZ01R01
Total lines 0012 ...............................STATEMENT DEFINITION .........................................
***                        TOP OF LIST
001  MOVE MAPNAM.FIELD TO ZZ01R01.KEYFIELD; load key for I/O
*** --------------------------------- PROCESS OPTION ---------------------------------
002  IF ZZ01R01 IS NRF         ;  If no record found
003    MOVE 'NO RECORD MATCHING KEY' TO EZEMSG;
004  ELSE;
005    IF ZZ01R01 IS ERR; -- unexpected I/O problem
006      PERFORM ZZ01P-WHOOPS;
007    END;
008    MOVE 'FOUND A WINNER!' TO EZEMSG;
009    MOVE ZZ01R01.FLD1 TO MAPNAM.FLDA    ; move indiv. record field to map
010    MOVE ZZ01R01 TO MAPNAM               ; move all like-named fields recd-map
011    PERFORM ZZ01P-CALCTAX;
012  END;
***                        END OF LIST
```

Figure 8-6. INQUIRY process statements.

Over the next several pages, the specification of various PRO-CESS OPTIONs is discussed. The PROCESS OPTIONs covered are all for direct processing of a record based upon its key. The record to be used must be defined as the OBJECT of the PROCESS OPTION. Here are the five PROCESS OPTIONS and their meanings:

Process Option	Function
INQUIRY	Read an individual record
UPDATE	Read a record and place a hold on it awaiting further REPLACE or DELETE processing
REPLACE	Rewrite a record that was previously read using UPDATE (or SCAN for update when a database is involved, see Chapter 12)
ADD	Create a new record
DELETE	Delete a record that was previously read using UPDATE (or SCAN for update when a database is involved, see Chapter 12)

Inquiry—Process Option

The INQUIRY process option is used to read an individual record based upon a key value. The key was defined as part of record definition and a value is moved into the key BEFORE execution of the process option. This is a logical use for the BEFORE part of the process definition.

INQUIRY (Figure 8.7) requires the name of an object to perform I-O against. Records will be returned using the format described in the record definition. Data in the record is available in the AFTER portion of the process logic.

APPLICATION PROCESS LIST			
SEL PROCESS	OPTION	OBJECT	ERROR
000 name_of_routine	Inquiry	record_name	EZERTN

Figure 8.7.   INQUIRY process option.

An ERROR routine should be named for all file/database I-O processes. If no ERROR routine is named, any non-zero response from the I-O will end the application's execution. Many developers use the special CSP word EZERTN to indicate that the process's AFTER logic will deal with any I-O irregularities. Some shops create common I-O error handling routines that pass "OK" or "NOT OK" switches of some kind.

Update—Process Option (Read for Update)

The UPDATE process option is used to read an individual record based upon a key value for the purpose of changing or deleting it. The key was defined as part of record definition and a value is moved into the key BEFORE execution of the process option. This is a logical use for the BEFORE part of the process definition.

APPLICATION PROCESS LIST			
SEL PROCESS	OPTION	OBJECT	ERROR
000 name_of_routine	update	record_name	EZERTN

Figure 8.8.   UPDATE process option.

UPDATE requires the name of an object to perform I-O against. Records will be returned using the format described in the record definition. Data in the record is available in the AFTER portion of the process logic. The UPDATE causes the record retrieved to be locked until the end of the application's execution or until a RE-PLACE or DELETE is done. The duration of locks is dependent upon data organization type and execution environment.

An ERROR routine should be named for all file/database I-O processes. If no ERROR routine is named, any non-zero response from the I-O will end the application's execution. Many developers use the special CSP word EZERTN to indicate that the process's AFTER logic will deal with any I-O irregularities. Some shops create common I-O error handling routines that pass "OK" or "NOT OK" switches of some kind.

Replace—Process Option

The REPLACE process option is used to modify an individual record that was read previously using the UPDATE process option. REPLACE is also valid when processing database records after a SETUPD and one or more SCANs.

REPLACE (Figure 8.9) requires the name of an object to perform I-O against. The record's data should be modified before issuing the REPLACE. In some environments, the REPLACE will free locks obtained by the UPDATE. Check with your CSP documentation for your system to be sure.

APPLICATION PROCESS LIST			
SEL PROCESS	OPTION	OBJECT	ERROR
000 name_of_routine	replace	record_name	EZERTN

Figure 8.9. REPLACE process option.

An ERROR routine should be named for all file/database I-O processes. If no ERROR routine is named, any non-zero response from the I-O will end the application's execution. Many developers use the special CSP word EZERTN to indicate that the process's AFTER logic will deal with any I-O irregularities. Some shops create common I-O error handling routines that pass "OK" or "NOT OK" switches of some kind.

Add—Process Option

The ADD process option is used to create new records based upon a key value. Record data and the key value should be moved into the record area BEFORE execution of the process option. ADD (Figure 8.10) requires the name of an object to perform I-O against. Records will be created using the format described in the record definition.

APPLICATION PROCESS LIST			
SEL PROCESS	OPTION	OBJECT	ERROR
000 name_of_routine	add	record_name	EZERTN

Figure 8.10.   ADD process option.

Some systems will return DUP or UNQ conditions (see Chapter 5) to warn of duplicate key conditions in the base file or an index. If this is an error, the record is not added.

An ERROR routine should be named for all file/database I-O processes. If no ERROR routine is named, any non-zero response from the I-O will end the application's execution. Many developers use the special CSP word EZERTN to indicate that the process's AFTER logic will deal with any I-O irregularities. Some shops create common I-O error handling routines that pass "OK" or "NOT OK" switches of some kind.

Delete—Process Option

The DELETE process option is used to eliminate an individual record that was read previously using the UPDATE process option. DELETE is also valid when processing database records after a SETUPD and one or more SCANs.

DELETE (Figure 8.11) requires the name of an object to perform I-O against. Record data should be backed up as necessary before the DELETE. In some environments, the DELETE will free locks obtained by the UPDATE. Check with your CSP documentation for your system to be sure.

APPLICATION PROCESS LIST			
SEL PROCESS	OPTION	OBJECT	ERROR
000 name_of_routine	delete	record_name	EZERTN

Figure 8.11.   DELETE process option.

An ERROR routine should be named for all file/database I-O processes. If no ERROR routine is named, any non-zero response from the I-O will end the application's execution. Many developers use the special CSP word EZERTN to indicate that the process's AFTER logic will deal with any I-O irregularities. Some shops create common I-O error handling routines that pass "OK" or "NOT OK" switches of some kind.

DATA INTEGRITY CONCERNS

In the UPDATE cycle on the previous pages, a problem common to online processing becomes evident: Two users might attempt to update the same record at the same time! A system developer is faced with two alternatives:

1. Lock all records that MIGHT be modified so that others cannot use them.
2. Do not lock records until necessary, *but then* you must account for the possibility that the record might have changed between the time you displayed it originally and the time the user completed their changes.

Locking all records that might get changed sounds good, but in practice it tends to cause contention problems. Locking all records is IMPOSSIBLE to implement when using segmented application processing under CICS or DPPX. The second method works only if some test can be devised to detect that a record has changed. Three common tests are the following:

1. Save the old record image and compare to new.
2. Save date and timestamp in record and store it between display and update for comparison reasons.
3. Store version numbers in record and compare.

In any case, it is important that the application design account for the eventuality (if it is likely). Decide either to disallow multiple access of the same record or to notify the user that the record they are attempting to change is no longer in the same form.

ACTIONS TAKEN TO DISPLAY THE CONTENTS OF A RECORD

Here (as a hint for the coming exercise) is the process required to display the contents of a record.

1. Define map (unless already defined).
2. Define record (unless already defined).
3. Clear map.
4. Clear record.
5. Move message to map requesting entry of record key (it is probably best to always use EZEMSG field name for the message field).
6. Use CONVERSE to display and receive.
7. Use key entered to issue INQUIRY.
8. Move appropriate data to screen.
9. CONVERSE.

%GET—COPYING CODE

One of the first things a good programmer or application developer learns to do is copy code. (Next you learn who *not* to copy *from!*) CSP provides a mechanism for copying code from one process or statement group to another. The "%GET" command allows you to copy Process or statement group statements from one Process or Statement group to another. %GET will copy from a named process/group into the portion of the receiving process/ group where the %GET is placed (see Figure 8.12).

%GET name

%GET appl.name

Figure 8.12.   Copying statements with %GET.

There is a simple two-step procedure used for copying code with %GET.

1. Determine the name of the process/group to be copied from.
2. Insert "%GET copyfromname" on a new line where the desired statements will go and press PF4.

The copied statements will be preceded by a comment indicating their source. This command is designed to copy like portions of code from one process/group to another. Using the %GET in the top portion of a process (BEFORE the I-O) copies only statements from the top (BEFORE) portion of the source. Using the %GET in the bottom portion of a process (AFTER the I-O) copies only statements from the bottom (AFTER) portion of the source. When copying into an EXECUTE process or statement group, only the top (BEFORE) portion of a source I-O process is copied. To get code from the bottom (AFTER) portion of a process into an EXECUTE process or statement group, it must first be moved into the top (BEFORE) portion of the target group. Use copy to create a duplicate of a potential source; then use the "M" or "MM" line commands to move the statements as desired. Delete the dummy source process when done.

CHAPTER 8 EXERCISES

Questions

1. How does CSP perform I-O?
2. The processes performing I-O are divided into two parts—why?
3. I-O processes require an object definition. What should be entered as the object for an I-O process?
4. Do all processes require the specification of ERROR routines?
5. What happens to an application when a non-zero return results from an I-O and no ERROR routine is specified?
6. What is a good use of the BEFORE process option logic in an INQUIRY or UPDATE process?
7. In what part of an INQUIRY process may you correctly move data from the record to a map?

8. What must occur before an UPDATE or DELETE process may be executed?
9. When does ERROR logic execute?
10. Why is data integrity a special concern in online systems?

Computer Exercise
Simple Record Processing

1. Change your application to access the record that matches the INVENTORY-ID entered and edited in a previous exercise.
2. Your application should once again display your map with a message prompting the user to enter a record key.
3. If the user keys in the word "QUIT" or presses PF12, end the application (maybe using EZECLOS?).
4. After receiving the key, your application should edit the key that was entered as done in a previous exercise (Chapter 5).
5. If the edits are okay, use the record specified in Chapter 6's exercise to attempt to obtain the record's data. If the edits fail, redisplay the map with error message.
6. If the record is found, move the appropriate data from the record to the map and display a success message.
7. If the record was not found, redisplay the map with error message.

Please read the pseudocode and error checking hints in Figure 8.13 before beginning.

Possible pseudocode:

```
Clear map & screen areas
Move message to map asking for inventory id
Do while user has not pressed PF3 or entered "quit"
      Display map
      Edit inventory id
            If inventory id OK
                  Read record
                  If record found
                        Move record data to map
                        Move message to map asking for next inventory id
                  Otherwise (no match on file)
                        Move error message to map
            Otherwise (inventory id failed edit)
                  Move error message to map
End Do
Exit Application
```

HINT: Don't forget error handling:

```
IF recname IS NRF              ;EZERT8 = 202,203,205,207
      PERFORM NOTFND;
ELSE ;
      IF recname NOT ERR        ;EZERT8 = 0
            PERFORM RECFND;
      ELSE                      ;EZERT8 NE 0
            PERFORM IOERR;
      END;
```

Figure 8.13.   Possible pseudocode.

9

Using the CSP
Message Facility

Okay, time for a breather. This chapter is quick and easy, but it introduces you to a nifty part of the CSP design. CSP's built-in error message facility is a well-designed tool. Please give its use some consideration.

Error messages used by online applications tend to fall into three stylistic groups:

Error message only displayed
Error code displayed
Combined error code plus message displayed

Many organizations have found the third option, combining error codes with messages, works best for system development. Using both error codes and messages means that the application developer need only refer to the error by some code. The error codes may be part of the system design or developed "on the fly," but in every case they should be recorded. Messages are then created for each error code as necessary. Many shops have found placing the actual creation of message text into user hands is a winning strategy. This removes the responsibility for the message text from the application development team without harming the

185

implementation effort. Since the application developer needs to know only the codes, it doesn't matter what the messages actually say. Of course, it's a good idea to make sure that messages and error codes are appropriate as part of the quality control process. (You know, walk throughs and all that stuff you normally do.)

CSP has a built-in message facility that is fast and simple to use. A VSAM RRDS (Relative Record Data Set) is defined to hold the messages. A CSP-provided utility is used to load and change the messages. An application simply uses EZEMSG as a field name; then, when a value is moved into EZEMNO, the system automatically inserts the appropriate message. Presto! Of course, it's not quite *that* simple, but almost.

The entire system has been designed into CSP as an integral part from the beginning. All you have to do is use it. This means that the message portion of the code doesn't have to be a maintenance worry; you deal with it only once per application. Before we discuss this in detail, let me review the basic steps that will be covered.

1. A VSAM RRDS of the appropriate format must be defined.
2. The CSP Message Utility application (or something that takes its place) is installed. IBM provides a sample application in source form that may be used as is or modified to fit your needs.
3. Messages are entered into the message file using the message facility.
4. Applications that will use the message file reference it as part of the application definition.
5. The message file is defined to the test and/or execution facilities.
6. Applications that will use the message facility include a field named EZEMSG on every map that may display messages.
7. Applications move message number values into the special field EZEMNO before a CONVERSE process sends a map.
8. The map displays the message associated with the message number.
9. The user is once again thrilled and amazed at the prowess of the data processing professional!

The rest of this chapter takes you step-by-step through the process of using the CSP Message Facility. Instructions are included to walk you through the process of using CSP's Message Utility for test purposes. In Chapter 14 you will find detailed instructions for installing the Message Utility more permanently. (Check first—it may already be installed.)

MESSAGE FILE FORMAT AND DEFINITION

The CSP message file must be a VSAM RRDS (Relative Record Data Set). The format of CSP's files sometimes changes from version to version, so be careful that you create the file required for your installation. The example in Figure 9.1 is for CSP Version 3 Release 3.

Name (message.dataset.name) may be any DSN valid for your environment. Since CSP deals almost exclusively in DDNAMEs, it is a good idea to follow a naming convention. Many installations have adopted a standard that the DDNAME used and the last node of the DSN is the same. This allows someone familiar with the standard to tell at a glance what file they are dealing with.

Size (CYL or TRK or REC) of the dataset may be whatever you think appropriate. Remember that VSAM will attempt 123 secondary allocations, so don't get too greedy. Also remember that your system may use MVS SMS (System Managed Storage) and the file size may be altered for you automatically.

Record size (RECSZ) must be 256 bytes. CSP's utility expects to find up to three messages per 256 byte record, but it expects that 256 byte record! Don't worry about the format—that's CSP's job. Volume number (VOL) will be assigned to you by a systems programmer in your environment.

```
DEFINE CLUSTER (  NAME      (message.dataset.name)    -
                  CYL       (n n)                      -
                  RECSZ     (256  256)                 -
                  VOL       (xxxxxx)                    -
                  NUMBERED)
```

Figure 9.1.   Message file—IDCAMS definition statement.

The last option (NUMBERED) tells VSAM to create a Relative Record Data Set.

If you are new to VSAM IDCAMS (Virtual Storage Access Method; IDC is IBM's product code for VSAM, Access Method Services) the syntax may be entered exactly as shown either in batch or from a TSO terminal at the READY prompt (or by using ISPF/PDF option 6, or by issuing the TSO command from a command line). The hyphens at the end of each line are continuation marks and are unnecessary if you can squeeze all of this onto one line.

CSP-PROVIDED MESSAGE FILE APPLICATION (MS10A)

Every CSP/AD system has a special read-only MSL available with the DDNAME of UTILMSL. In this MSL are several sample applications supplied by IBM as part of the CSP/AD package. If your installation is especially tight on disk space, this MSL might have been dropped from the system. Check with your system administrator if you cannot find the MS10A application in an MSL called UTILMSL.

UTILMSL should contain an application named MS10A. This application is the Message File Utility provided by CSP in source code form and may be modified to fit your needs and standards. You are, however, stuck with the message number format as designed by CSP's developers.

Before using MS10A, the message file must be allocated to your CSP/AD session. If you are using CSP/AD under CICS, this means that an FCT entry is required for your message file pointing to the DDNAME "MS10F01." If you are using TSO, the allocate statement in Figure 9.2 is required.

ALLOC DD(ms10f01) DSN('message.data.set') SHR REU

Figure 9.2.   Message file allocation for MS10A.

CSP's Message File Utility (MS10A) expects to find a file with the DDNAME MS10F01. If this is inconvenient, simply alter the record definition in the MSL to point to the FILE name you de-

```
EZEM01                        CROSS SYSTEM PRODUCT

==>
                    ENTER = Continue      PF3 = File and Exit

..................................... MSL SELECTION ...............................................

         MSL FILE NAME          MSL ACCESS          CONNECT
                                  ORDER              STATUS

Read/Write MSL:
      =>  yourmsl                => 1

Read-Only MSLs:
      =>  utilmsl               => 2
      =>                        => 3
      =>                        => 4
      =>                        => 5
      =>                        => 6


                        In all facilities:
                  PF1 = Help     PA2 = Cancel
```

Figure 9.3.   Adding UTILMSL to MSL concatenation.

sire. Anyway, the file name must be known to CSP before testing the Message File Utility.

Next, add UTILMSL to your current MSL concatenation. In most systems this file will have been allocated automatically as part of CSP/AD, and you will not have to do anything special to allocate it. See Figure 9.3.

Enter the CSP/AD TEST facility to begin use of the Message Utility. The example in Figure 9.4 shows a configuration of TEST's initial panel that I have found successful. Note that the TRACE has been turned off, the number of statements to execute has been increased, and a STOP key has been specified. None of these things is required, but it seems senseless to trace an application that works or to stop unnecessarily as it executes.

The next panel that displays should be the MESSAGE FILE MAINTENANCE panel (unless you renamed it as part of your customization). See Figure 9.5. Key in the message number you wish to create, view, or change. Message numbers must be nu-

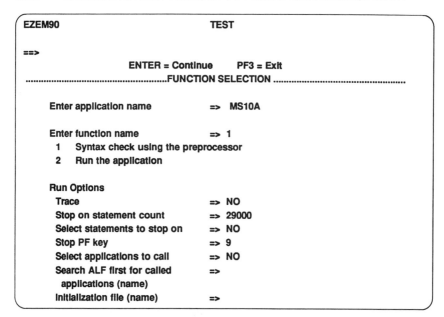

Figure 9.4.   Using CSP TEST to run MS10A.

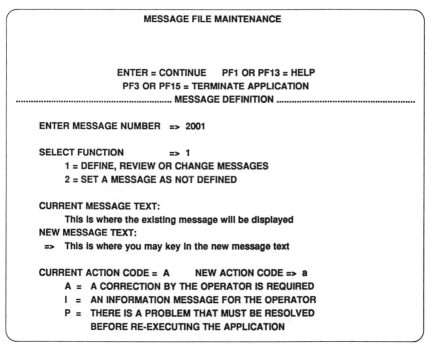

Figure 9.5.   MS10A message file maintenance panel.

meric, up to four (4) digits long. It is a good idea to standardize the message numbers in a system. Some installations reserve blocks of numbers (say 0300–0399) for a particular application.

Select a function; "1" means to create, review, change a message and "2" is used to delete (sort of) a message. Since the file is a VSAM RRDS, deletion is really a matter of flagging a record as "not here" rather than actually eliminating it. MS10A defaults to "1," which is great, since that's what we normally use anyway. The current message text will display with an entry area for the new message text. Enter the new message (up to 68 characters) text as you would like to see it appear.

You may wish to fiddle with CURRENT ACTION CODE. MS10A's designers thought that it would be a good idea to have the message number show the relative severity of the message. ACTION CODE allows you to do just that. MS10A's designers set up three classifications: "A" telling the user they made a mistake, which must be corrected; "I" providing some useful information; and "P" indicating that a system problem exists. Many shops have modified the MS10A application to allow more or different codes.

After filling in the screen's fields as desired, press ENTER to create, view, or change the message. Pressing ENTER repetitively causes MS10A to "walk" through the message file one message at a time. You may also enter the next desired message number to go directly to a specific message.

The error message displayed by CSP's message facility has a specific format.

1. The first three characters match part of the message file ddname (DCAxxxD). The ddname format is dictated by system design, the fourth, fifth, and sixth characters are added to each application's definition that will use the file.
2. The fourth character is an optional value added to each application's definition after the three ddname characters. If unused, this displays as a blank.
3. The fifth through eighth characters are four digits containing the message number. Leading zeroes are displayed if applicable.
4. The ninth character is the ACTION character defined to the MS10A application.

5. The tenth character is a single blank (always).
6. The following characters (up to 68) represent the message text. The EZEMSG field must be large enough to hold the message number (as displayed by CSP) and the message.

ADDING THE MESSAGE FACILITY TO AN APPLICATION

To use the built-in Message Facility in an application, three things must be done: The message file must be defined to the application, a field named EZEMSG must be defined, and the EZEMNO field must be loaded with appropriate values prior to CONVERSEs.

The APPLICATION SPECIFICATIONS panel has a place for defining the message file to be used (Figure 9.6). The first three characters of the Message File entry must match the fourth, fifth, and sixth characters of the file's ddname. The ddname format is dictated by CSP's design to be DCAxxxD. The "DCA" at

```
 _____
/                                                                   \
|  EZEM31                      APPLICATION DEFINITION                 |
|                                                                    |
|  ==>                                                               |
|                      PF3 = Exit (or continue if new definition)    |
|                      Application Name = QQ01A                       |
|  ....................................... APPLICATION SPECIFICATIONS .......................................  |
|                                                                    |
|                                                                    |
|      Type of application => 1                                      |
|                                     Working Storage  = QQ01W01      |
|          1   Main Transaction                                      |
|          2   Main Batch             Map Group Name => QQ01G         |
|          3   Called Transaction                                    |
|          4   Called Batch           Message File    => xxxA         |
|                                                                    |
|      Help Map Group Name =>         PSB Name         =>             |
|                                                                    |
|      Help PF Key         =>         PF1-12=PF13-24   => YES         |
|                                                                    |
|      Bypass Edit PF Keys  => 3      Allow Implicits  => NO          |
|                                                                    |
|                                     First Map        =>             |
_____/
```

Figure 9.6. Specifying message file during application definition.

the beginning and the "D" at the end are required. The "xxx" may be replaced by any three characters that "tickle your bippy" (hopefully, a more scientific method is used for naming files, but whatever works). The three characters that you choose are the same three characters that must be entered in the APPLICA-TION SPECIFICATIONS panel's Message File field. CSP provides room to add a fourth (optional) character. The intent of this fourth character is to provide the ability to tell which application a message came from. If two or more applications use the same panel and might generate the same message number, it might be difficult for a maintenance person to figure out which CSP application to work on. If you use different fourth characters for the message file when defining it to different applications, then, when the message appears, you will know where it came from.

To use the built-in facility, maps must define a field named "EZEMSG" somewhere. The "EZEMSG" field must be long enough to show the message number (all ten bytes) plus enough message to be meaningful. A field length greater than 78 is wasteful since the message can't be any longer.

```
EZEM39                    APPLICATION PROCESS DEFINITION

==>
                    PF3 = File and exit  (or file and continue if more selected)
            Process => QQ01P-GETREC          Description = Get a record
            Option = INQUIRY                 Object  = QQ01R01
Total lines 0012 ...........................STATEMENT DEFINITION .........................................
***                          TOP OF LIST
001  MOVE MAPNAM.FIELD TO QQ01R01.KEYFIELD; load key for I/O
*** ------------------------------- PROCESS OPTION -----------------------
002  IF QQ01R01 IS NRF          ;   if no record found
003      MOVE 219 TO EZEMNO;
004  ELSE;
005      IF QQ01R01 IS ERR; -- unexpected I/O problem
006          PERFORM QQ01P-WHOOPS;
007      END;
008      MOVE 220 TO EZEMNO;
009      MOVE QQ01R01.FLD1 TO MAPNAM.FLDA   ; move indiv. record field to map
010      MOVE QQ01R01 TO MAPNAM             ; move all like-named fields recd-map
011      PERFORM QQ01P-CALCTAX;
012  END;
***                          END OF LIST
```

Figure 9.7. Using EZEMNO to control messages.

To cause CSP to display a message, move a message number into the special CSP field "EZEMNO" before a CONVERSE or DISPLAY (Figure 9.7). When CSP sends a map to the terminal, it will use the current value in EZEMNO to decide which message to display. CSP automatically searches the message file for a matching message and displays it. If no message is found, the message number displays with a message indicating that the message could not be found. If EZEMNO equals zero, CSP will not alter the EZEMSG field.

That's all there is to it. CSP's message facility is efficient, easy to use, and maintenance free. What more could you ask? (Don't the words "transparent to the user" chill you to the bone?) Honestly, this is a good facility and you should use it.

CHAPTER 9 EXERCISES

Questions

1. What type of dataset is created to store CSP messages?
2. What must the ddname be in order to use the Message File Utility (MS10A)?
3. What MSL is the source code for MS10A found in?
4. Can you change the format of the CSP error message?
5. What must the ddname be in order to use the Message File from an application?
6. What part of a Message File's ddname is entered on the APPLICATION SPECIFICATIONS panel?
7. What field must be defined in a map in order to use the CSP message facility?
8. What field do you move a message number into in order to use the message facility?
9. How can you cause CSP to leave EZEMSG alone?
10. Is using CSP's message facility a good thing?

Computer Exercise
Message Creation and Use

1. Locate a message file or create one for your own use.
2. Allocate the file for use by the message utility.

3. Execute CSP's message utility (MS10A) and add messages for the following:

   Inventory ID invalid

   Inventory ID not found, try another

   Inventory record displayed as requested

   Unexpected error, munch antacids and call for help

4. Modify the application you have been working on to display your newly created error messages from the message file.
5. Allocate the message file for use by your application.
6. Test your application and see if the messages appear as they should.

10

Data Tables

Many programming tools provide the capability of array processing to make processing simpler and sometimes more efficient. While creating CSP records, you may use the OCCURS column to identify an array within a record. CSP also gives you the capability of creating arrays outside of record definitions called CSP DATA TABLES. CSP data tables are stored and processed in virtual memory, making them a fast, easy-to-use tool. CSP data tables are not the same as SQL tables, or Dialog Manager tables, or CICS data tables. CSP data tables are useful only to CSP applications.

Data stored in CSP data tables may be used as an array in much the same way as third-generation languages processed arrays. In addition, CSP can use the data tables in automatic map edits. Data tables are generated separately from maps and applications. Data tables may be shared by multiple applications in some environments, thus increasing efficiency by decreasing memory requirements.

CSP data tables are defined using a series of panels in the same way that maps, applications, and records are defined. As part of the design process, remember that CSP provides a table definition and access facility for two purposes:

Editing (via CSP's built-in map edits)
Reference (via CSP array-handling statements)

The same table may be used for both edit purposes and reference; your design dictates the need. An edit table is named when defining the map on the VARIABLE FIELD EDIT DEFINITION screen. When an edit table is specified, CSP takes care of testing the value(s) and sending a message back to the user (you specify the error message or CSP uses default messages). Edit tables may be used for three basic tests:

Is the entered value in the table?
Is the entered value *not* in table?
Is the entered value in a range included in a table row?

Tables may also be referenced by simply subscripting a reference to a table data item or by using special CSP table-handling statements. CSP's IF . . . IN, FIND, RETR, and MOVEA statements are all used for accessing data tables.

DEFINING CSP DATA TABLES

CSP tables are defined via a series of panels in the same manner that other CSP objects are defined. The first time a table is defined, CSP/AD will automatically move through part of the definition as you press the PF3 (exit) key. To complete the definition, and any time you modify the table, the FUNCTION SELECTION panel is where you start.

To create a data table, you start at the CSP DEFINITION SELECTION panel. Choose option 2 from the CSP/AD FACILITY SELECTION "main menu" panel or issue the =Edit command to display the DEFINITION SELECTION panel (Figure 10.1).

Table names may be one to seven (1–7) characters long. The first character of a table name must be a letter (A–Z). Other characters in a table name may be any combination of numbers (0–9) and letters (A–Z); special characters are not allowed. The first three characters may not be "EZE," and the last character may not be the number zero (0). Specify member type 2 (Table) and press ENTER to continue.

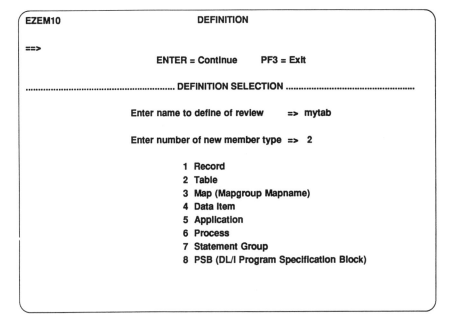

```
EZEM10                           DEFINITION

==>
                        ENTER = Continue      PF3 = Exit

.................................... DEFINITION SELECTION ....................................

            Enter name to define of review      => mytab

            Enter number of new member type  =>  2

                        1  Record
                        2  Table
                        3  Map (Mapgroup Mapname)
                        4  Data Item
                        5  Application
                        6  Process
                        7  Statement Group
                        8  PSB (DL/I Program Specification Block)
```

Figure 10.1.   Definition selection.

```
EZEM51                        TABLE DEFINITION

==>
                  ENTER = Continue              PF3 = File and exit
                        Table Name = mytab
.................................... FUNCTION SELECTION ....................................


            Enter function number => 1

                1      Table Specification
                2      Column Definition
                3      Contents Definition
                4      Table Prologue
```

Figure 10.2.   Table definition—Function selection.

The TABLE DEFINITION FUNCTION SELECTION panel (Figure 10.2) does not display the first time you create a table. To fill in table contents, or to make table modifications, this table is where you start. The choices listed are the following:

1. Table Specification, to define the basic table properties.
2. Column Definition, to define the table's data format.
3. Contents Definition, to load data into a table.
4. Table Prologue, to document the table. (There's that "D" word again!).

The Table Specification panel (Figure 10.3) is used to tell CSP what type of processing is anticipated using the table. You may only choose one of the automatic edit styles. If the same table data might be useful for an automatic match valid edit and an automatic match invalid edit, two tables must be defined.

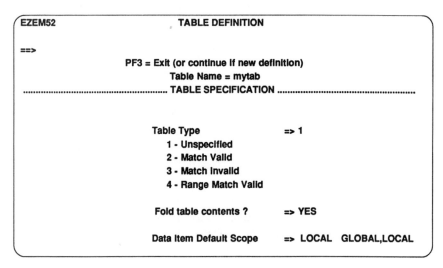

Figure 10.3. Table definition—Table specification.

The choices for table type are listed on the panel. All tables may be used by CSP's table-handling statements. For a table to be used by an automatic edit, a table type must be chosen:

1. Unspecified—this table is for reference only and will not be used by automatic map edits.
2. Match Valid—if specified for automatic map editing, the data entered must match a value found in the first column of the table.

3. Match Invalid—if specified for automatic map editing, the data entered must not match any value found in the first column of the table.

4. Range Match Valid—if specified for automatic map editing, the data entered must fall between (inclusive) the first column's value and the second column's value of some row in the table.

Folding (Fold => YES) indicates that when data is entered for the table, lower-case letters are to be converted to upper-case upon entry or regeneration. Data Item Default Scope means the same thing here that it does in record definition. GLOBAL means that an entry in the current MSL list will be used for data item definitions if available, or that an entry will be created in the read-write MSL. LOCAL means that the table's data items are to be part of the table definition only and will not be represented by a member in the MSL. Once the table's properties have been defined, its columns must be described. The TABLE DEFINITION COLUMN DEFINITION panel is similar to the RECORD DEFINITION DATA ITEM DEFINITION panel. Column names and attributes are assigned.

The table in Figure 10.4 shows two columns being defined. Columns are data items in CSP, and all normal data item rules apply to their naming and definition with two exceptions. Column data items may be no larger than 254 bytes long, and column data items may not OCCUR (CSP does not support two-dimensional tables).

```
 EZEM53                        TABLE DEFINITION

 ==>
                                PF3 = Exit
                             Table Name = mytab
 Total lines 0002 ................................. COLUMN DEFINITION ...........................................
       NAME                          LEVEL  TYPE   LENGTH DEC   BYTES  SCOPE
 ***                        TOP OF LIST
 001 STCODE                           10    CHA    00002        00002  LOCAL
 002 STNAME                           10    CHA    00015        00015  LOCAL
 ***                        END OF LIST
```

Figure 10.4.   Table definition—Column definition.

```
EZEM54                          TABLE DEFINITION

==>
                                PF3 = Exit
Total positions 00102           Table Name = mytab          Positions 00001 to 00102
Total lines   0004 ........................ CONTENTS DEFINITION .................................................
***     STCODE    STNAME
001     AZ        ARIZONA
002     CA        CALIFORNIA
003     CO        COLORADO
004     FL        FLORIDA
005     NH        NEW HAMPSHIRE
006     TX        TEXAS
***     STCODE    STNAME
```

Figure 10.5.   Table definition—Contents definition.

Once the columns have been defined, exit back to the TABLE DEFINITION FUNCTION SELECTION panel.

If you would like to load the data table with information, choose option 3 (Contents Definition) from the TABLE DEFINITION FUNCTION SELECTION panel and press ENTER. On the panel in Figure 10.5 you may enter rows of data to be stored in the CSP data table for use by applications. Each line in the input area represents one row of table data. CSP allows up to 4092 (inclusive) rows in a table. CSP tables are always searched sequentially from top to bottom. Place the most frequently used data first in the table to speed processing. The only way to alter table contents is in CSP/AD. Any changes to data tables during application execution will be erased the next time the table is loaded. Any table that is referenced by an application must be referenced in the application's ADDITIONAL TABLES AND RECORDS panel (Figure 10.6).

Reach the ADDITIONAL TABLES AND RECORDS panel by using option 3 from the APPLICATION DEFINITION FUNCTION SELECTION panel. Enter the table and the code to show that it is a table rather than a record. If you do not list the table on this panel, the application and its processes will not be able to "see" the table and error messages will result.

```
┌─────────────────────────────────────────────────────────────────────────┐
│ EZEM35                        APPLICATION DEFINITION                     │
│                                                                           │
│ ==>                                                                       │
│                                 PF3 = Exit                                │
│                           Application Name = myappl                       │
│ Total lines 0001 ................ TABLE AND ADDITIONAL RECORDS LIST .....................................│
│          MEMBER NAME                        TYPE:      1 Table    2 Record │
│ ***                                    TOP OF LIST                        │
│ 001    mytab                                  1                           │
│ ***                                    END OF LIST                        │
│                                                                           │
│                                                                           │
└─────────────────────────────────────────────────────────────────────────┘
```

Figure 10.6. Table and additional records list.

SHARING TABLES

During the GENERATION process (see Chapter 14) for a table, CSP will ask if the table is to be SHARED. In some environments data tables may be SHARED—that is, only one copy of the table is kept in memory no matter how many users will share it. This can provide a significant savings in environments like CICS. Marking "SHARE = YES" during table generation enables a table to be generated as shared. Tables may be shared on CICS, IMS/TM ESA (IMS-DC), and DPPX systems (but not using IMS BMP).

Under CICS, the table *must* be marked as RESIDENT using the ALF utility (more on both of those in Chapter 16) in order for sharing to happen. Under VM and TSO, "SHARED = YES" does *not* allow sharing, but it is required for tables that will be marked as RESIDENT.

UPDATING TABLE DATA

Table data may be modified by applications that use the table. However, table data is lost when the table is deleted from memory and reloaded based upon the definition the next time the table is used. The "life" of changed data is dependent upon when the table is deleted. This is mostly controlled by three things:

KEEP AFTER USE = YES (generation option)
RESIDENT / NONRESIDENT (controlled with ALF utility)
execution platform

If a table is RESIDENT (not unloaded when done executing):

Under TSO & CMS the table is deleted at the end of the main
    application.
With CICS the table is deleted only by ALFUTIL NEW COPY
    (or end of the CICS address space).

If a table is NONRESIDENT (all execution platforms):

KEEP AFTER USE = NO means that the table is deleted at end
    of the main process that caused it to be loaded.
KEEP AFTER USE = YES means that the table is deleted at the
    end of the applications that use it.

Under COBOL execution environments, KEEP AFTER USE is
ignored, and tables are kept until the end of the application (segment). Other restrictions apply for DPPX and OS/400 systems.
Check your system documentation.

USING TABLES IN AUTOMATIC EDITS

One of the more popular ways to use CSP data tables is as part of
the built-in edits provided by CSP. Using automatic edits has
both positive and negative points. First the good stuff: If you use
CSP to automatically search the tables looking for match, no
match, or range match conditions, you don't have to write the
code! Nor do you have to update the code as CSP versions change.
The downside is significant *sometimes*. If you have lots of users
and they make lots of mistakes, the extra line transmission
caused by the automatic edit routines can chew up lots of response time. This is because a system doing manual edits might
do several field edits at once and return the screen to the user
with multiple fields highlighted looking for corrections. The act
of going back and forth one error at a time (like CSP's automatic
edits do) can cause excessive line transmissions. Again, this is

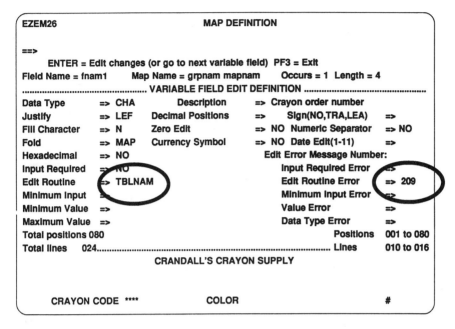

Figure 10.7. Naming tables for map edits.

normally only a problem if you have many users making many mistakes.

Name the edit table on the MAP DEFINITION VARIABLE FIELD EDIT DEFINITION panel (Figure 10.7). The table name goes in the field marked "Edit Routine =>" on the left side of the panel. The error message associated with a failed search is created by placing an error number in the "Edit Routine Error =>" position. Notice that the type of edit is not specified here; it was specified at the time of table definition! You just have to know what type of table you are using (tsk, tsk, probably requires more documentation).

TABLE PROCESSING STATEMENTS

CSP's statements provide several ways to access table data. Any table may be accessed using CSP statements, regardless of table type. FIND is used to get the first row in a table matching a search value. The IF . . . IN combination may be used to search a table looking for one or more matches. RETR searches a table and automatically moves data when it finds a match. MOVEA allows movement of all or partial tables. And finally, subscripting may be used just about anywhere.

FIND is used to find a data value in a table and then execute a chosen Statement Group when a match is found or to execute a Statement Group when a match is not found (or both). See Figure 10.8.

FIND itemval mytab[.colnam] fndrtn,notfnd

Figure 10.8. FIND statement syntax.

- itemval is a data item or literal to be searched for in the table.
- mytab refers to a CSP table definition.
- .colnam, if used, names the column in the table to be compared to itemval (the first column in the table is the default).
- fndrtn names the statement group or EZE routine to be executed if a match is found.
- notfnd names the statement group or EZE routine to be executed if no match is found for itemval.

If the FIND is executed as part of a FLOW, both the "found" routine and the "not found" routine may be "main" processes (this would be highly abnormal).

If FIND gets a hit (finds a match), EZETST will contain the row number of the first row that matched itemval (FIND is a sequential search of the table). If no match is found, EZETST will be set to zero.

Figures 10.9 and 10.10 search the table MYTAB looking for an STCODE value matching the value of the data item MAPST. If a match is found, GOTHIT executes and EZETST is set to a value indicating the first row in MYTAB where an STCODE from the

---

FIND MAPST MYTAB.STCODE GOTHIT,NOHIT;

FIND MAPST MYTAB GOTHIT,NOHIT;

Figure 10.9. FIND examples.

table matches the MAPST variable. If no match is found, the NOHIT routine would be executed and the value in EZETST will be zero.

Another way to search for table matches is with the IN comparator of the IF statement. This is a little more readable than using FIND and has the added benefit of choosing a starting point. So, with the IF . . . IN combination you can see if multiple matches occur in the table.

IF itemval IN mytab   **or**
IF itemval IN mytab(startsub)

Figure 10.10.   IF . . . IN statement examples.

- itemval is a data item or literal to be searched for in the array.
- mytab refers to a CSP table column or other subscripted data item.
- startsub (optional) controls which row in the array the search is to begin at; if startsub is omitted the search begins with the first row.

If a match is found, EZETST will contain the subscript of the matching row. If no match is found, EZETST will contain zero. Using IN rather than FIND allows processing of multiple occurrences of a value and searching any array, not just tables. The example in Figure 10.11 searches the table MYTAB looking for an STCODE value matching the value of the data item MAPST. If a match is found, GOTHIT executes and EZETST is set to a value indicating the first row in MYTAB where an STCODE matches MAPST. If no match is found, the NOHIT routine would be executed and the value in EZETST will be zero.

```
MOVE 1 TO MYSUB                      ;1st subscript
IF MAPST IN MYTAB.STCODE(MYSUB)      ;search table
    GOTHIT                           ;stmt group
ELSE;
    NOHIT                            ;stmt group
END;
```

Figure 10.11.   Sample table search using subscripts.

Another CSP statement specifically geared to data tables is RETR. RETR is used to find data in a table matching some outside value, then move corresponding table data into another field (Figure 10.12).

RETR itemval mytab[.colnam] retnitem [retcol]

Figure 10.12. RETR statement syntax.

- itemval is a data item or literal to be searched for in the table.
- mytab refers to a CSP table definition.
- colnam, if used, names the column in the table to be compared to itemval (the first column in the table is the default).
- retnitem names the data item to be modified by the RETR statement.
- retcol, if used, names the column in the table to be moved into retnitem when a match is found (the second column in the table is the default; do *not* qualify retcol).

If RETR gets a hit, EZETST will contain the row number of the first row that matched itemval (RETR uses a sequential search of the table). If no match is found, EZETST will be set to zero. Both examples below find the first row from the table "mytab" with a STCODE equal to the MAPST variable and move the corresponding STNAME into the MSTNAM variable. EZETST is then set to the subscript value of the row containing the match (if no match, EZETST will equal zero and MSTNAM is unchanged). See Figure 10.13.

RETR MAPST MYTAB.STCODE MSTNAM STNAME;

RETR MAPST MYTAB MSTNAM;

Search the MYTAB table looking for the first row where the MYTAB table's STCODE column matches the MAPST variable – then move the STNAME column from that MYTAB row into the MSTNAM variable

Figure 10.13. RETR examples.

```
IF mytab.col1(1) = mapin.field1 ...

MOVE 'sample data' to mytab.col2(34)

MOVE 'more stuff' to mytab.col3(subscpt)

MOVEA mytab2 TO mytab3

MOVEA 'yet more' to mytab4 for 12
```

Figure 10.14.   Manipulating table data.

And (of course) there's always subscripting. Data in CSP data tables may always be retrieved by the simple mechanism of subscripting. An occurrence number is enclosed in parentheses directly following an array/table data item. The occurrence number may be a numeric literal or a variable (Figure 10.14).

Finally, we once again see MOVEA. As mentioned in Chapter 5, you either love this thing or you hate it. MOVEA may be used with data tables as well as "normal" arrays. Figure 10.15 repeats the examples you saw in Chapter 5.

TABLE EFFICIENCY

Since they are stored in virtual memory, tables are relatively fast if they do not get loaded/unloaded too often. Frequently accessed tables should be considered good candidates for RESIDENT status. Be sure to use SHARED tables under CICS and IMS. CSP searches tables *sequentially*. Try to load data into tables in the order most commonly searched. Use direct references to subscripted items when available rather than using table handling commands.

Some people will use a CSP data table to replace searches of VSAM files or databases to convert codes into text. Other installations use data tables to temporarily store the results of SQL cursor processing to avoid reopening the cursor (this is a great idea if well thought out, and if the data may be "old").

```
MOVEA 'x' TO OUTARRAY;

   OUTARRAY (before)        0 1 2 3 4
   OUTARRAY (after)         x x x x x
```

```
MOVEA 'x' TO OUTARRAY(2) FOR 2;

   OUTARRAY (before)        0 1 2 3 4
   OUTARRAY (after)         0 x x 3 4
```

```
MOVEA INARRAY(3) TO OUTARRAY(2);
   INARRAY                  a b c d e
   OUTARRAY (before)        0 1 2 3 4
   OUTARRAY (after)         0 c d e 4
```

```
MOVEA INARRAY(3) TO OUTARRAY(2) FOR 2;
   INARRAY                  a b c d e
   OUTARRAY (before)        0 1 2 3 4
   OUTARRAY (after)         0 c d 3 4
```

- **NOTE: EZETST will contain the subscript of the last value changed by the command**

Figure 10.15. MOVEA examples.

CHAPTER 10 EXERCISES

Questions

1. Are CSP data tables accessible with SQL?
2. What is the maximum width of a data table column?
3. How many rows may a data table include?
4. What are the rules for a data table name?
5. Why is a table better than many types of file I-O?
6. What command(s) can be used to find out if a data value is in a data table?

7. Write the command to use the RETR command to search the first column in a table called STORIES for a match to the variable BEDTIME. When a match is found, move the information from the table's FUN column into the variable named ARE.
8. How would you find the first and second occurrence of the value RUBBER-DUCKIE in the table BATHTUB, using the subscript variable SING-A-LONG?
9. What variable is used by CSP after most table-handling commands to report the occurrence number of a match?
10. How should data be loaded into a data table for efficiency purposes?

Computer Exercise
Table Creation for Edits and Look-Up

1. Add a display-only field to the map you've been using called CLASS. Make it 40 characters long.
2. DEWEY-NBR (a field in the record you've been using) is a numeric field holding the general Dewey Decimal value associated with each book. Your job is to create a table converting the whole number part of the DEWEY-NBR to a character classification as follows:

DEWEY-NBR	CLASSIFICATION
000–090	Generalities
100–190	Philosophy and related disciplines
200–290	Religion
300–390	Social Science
400–490	Language
500–590	Pure sciences
600–690	Technology (applied sciences)
700–790	The arts
800–890	Literature
900–990	Geography and history

3. Create a table named BB01T01 to hold the above data.
4. Add statements to your application that will, after a record is read successfully, convert the record's DEWEY-NBR to a CLASSIFICATION and move the CLASSIFICATION to the new CLASS field defined on the map.

VSAM "Browsing"

So far, the file processing discussed involved just one record at a time. The process options discussed so far have all dealt with direct processing. The application might use INQUIRY to read, UPDATE to read for update, REPLACE to rewrite, ADD to insert, and DELETE to eliminate a record. In many applications it is useful to display groups of records and to allow the user to "scroll" through the file. The ability to "scroll" through the data is often called "browsing" and depends on the application's ability to deal with multiple records at once. CSP has process options geared specifically to sequential processing of data. The SCAN and SCANBACK process options provide the ability to read sequentially toward the end of the file (SCAN) or to read sequentially toward the beginning of the file (SCANBACK). SCANBACK may only be used with CSP INDEXED records representing VSAM KSDS.

MULTIPLE RECORD DISPLAYS

Many applications require the display of multiple records, Figure 11.1 illustrates a typical screen layout. To define a map like Figure 11.1, start by defining one line of data on the map. Set up the fields as you would like the top row of the table to be displayed (Figure 11.2).

CRANDALL'S CRAYON COMPANY - INVENTORY STATUS

CRAYON CODE	DESCRIPTION	QTY ON HAND	QTY ON ORDER
C111	UGLY PINK	988	100
C121	SUMMER YELLOW	144	0
C122	MAZE	801	0
C131	EVERGREEN	50	150
C132	OLIVE DRAB	434	0
C133	FRANGO MINT	133	950
C141	ADOBE	100	100
C142	JUST PLAIN BEIGE	455	350

Figure 11.1. Sample multi-record panel.

```
EZEM22                          MAP DEFINITION

==>
                     PF3 = Exit (or continue if new definition)
Total positions 080           Map Name = grpnam mapnam          Positions 001 to 080
Total lines     024 ...           C(#)  V(−)  S(/)              ... Lines 001 to 017
........................................................................................

/#CRANDALL'S CRAYON COMPANY#/


     ¬    #          ¬                    #   ¬   #        ¬   #
```

Figure 11.2. Multi-record panel—Field definition.

Use the ATTRIBUTE command to set the attributes for all fields on the line one at a time. If you will be defining CSP edit or formatting data for the fields, go to the EDIT DEFINITION panel and set them up, too. Next use the REPEAT command to make the number of occurrences you desire. Key in "REPEAT 7" on the command line and place the cursor on the line to be repeated, then press ENTER. The screen now has eight (8) lines, each with the attributes, definition, and edit settings of the first line. Now, go back and add column headings for appearance.

Define the rest of the map fields by REPEATing the first line (it generally works best to define the data columns first, then go back and add the column headings). Don't forget that the total number of lines that appear will be the repetition number plus one (for example "REPEAT 7" results in eight total lines). You might also wish to add blank lines between the repeated lines for spacing purposes.

To name the fields, you must go to the MAP DEFINITION VARIABLE FIELD NAMING panel (Figure 11.3). Subscript the field names (unfortunately, one at a time). Near the top and on the left side of the MAP DEFINITION VARIABLE FIELD NAM-

```
EZEM24                        MAP DEFINITION

==>
                   PF3 = Exit (or continue if new definition)
0001 <= Number of first field to name          Map Name = grpnam mapnam
.......................................... VARIABLE FIELD NAMING ....................................................
        NAME
     1  CCODE(1)
     2  CDESC(1)
     3  CQHAND(1)
     4  CQORDER(1)
     5  CCODE(2)
     6  CDESC(2)
     7  CQHAND(2)
     8  CQORDER(2)
Total positions 080                            Positions    001 to 080
Total lines   024 .....................................................................Lines        001 to 010

     CRAYON CODE  DESCRIPTION      QTY ON HAND   QTY ON ORDER
     ------------------------------------------------------------
        1            2               3             4
        5            6               7             8
```

Figure 11.3. Multi-record panel—Field naming.

ING panel is a field marked " <= Number of first field to name." Change the value in this field to scroll the list of names so that more may be entered. The application may now reference the field names using subscripts to place data in a particular row.

WORKING WITH MULTIPLE RECORD DISPLAYS

The application logic that loads each row on a multiple row display should probably be performed from inside a WHILE loop. Loop until the screen is full (you'll need a counter) or until you hit the end of the available data (whichever comes first). When performing browse operations using indexed or database information, a program is faced with two major alternatives:

1. Read all data eligible for browse, and store for later display
2. Read data only as needed

Reading all eligible data into a work file is sometimes desirable when a user will always be viewing/printing the entire group. Another time is when an SQL query will cause the creation of a large temporary table and subsequent "paging" or "browsing" by the user will cause that temporary table to be recreated repeatedly. This method works only if it is okay to have (slightly) old data on the screen.

Reading the data only as needed is usually the preferred condition when the user might look at one or two screens and go away, or if the user requires the most up-to-date information available. SQL queries that do not create temporary tables will be satisfied one row at a time by the database. Therefore, there is no advantage to processing and storing more records than can be viewed at once. The subject of SQL and temporary tables is not within the scope of this book. Check with your local DBAs for information about a specific case.

USING FUNCTION KEYS TO CONTROL SCROLLING

Perhaps the most common way to control scrolling is with function keys. By now, most computer users have encountered an application that allowed them to "scroll" from one page to another using function keys. The "traditional" keys used to control scrolling are PF7 (backward/up) and PF8 (forward/down). PF7 and PF8 are also described as the standard scrolling keys by IBM's SAA-CUA (System Application Architecture—Common User Access) standard. Following this standard is a good idea, and you'll be in good company.

STORING RECORD KEYS BETWEEN ITERATIONS

When displaying multiple pages and building them one at a time, it is essential that the program know which record should be used to begin the next page. This is done easily by storing the keys of two records, the first record being displayed on the screen and the last between iterations. By storing the first and last record keys, a program may use those keys to begin the sequential access in the next program iteration.

In the example in Figure 11.4, if Crayon Code is the record key, you would store "C111" and "C142" so the next iteration would have a starting point when scrolling backward or forward. Define two variables in WORKING STORAGE to store these keys. Since WORKING STORAGE is available for the entire duration of the application's execution (even when running segmented under CICS), the data will be available for the next iteration.

The technique is simple. For "forward" browsing, store the key of the first record read for future "backward" browsing, and as records are read, store the current key for continued "forward"

CRANDALL'S CRAYON COMPANY - INVENTORY STATUS

CRAYON CODE	DESCRIPTION	QTY ON HAND	QTY ON ORDER
C111	UGLY PINK	988	100
C121	SUMMER YELLOW	144	0
C122	MAZE	801	0
C131	EVERGREEN	50	150
C132	OLIVE DRAB	434	0
C133	FRANGO MINT	133	950
C141	ADOBE	100	100
C142	JUST PLAIN BEIGE	455	350

Figure 11.4.   Multi-record panel—sample.

browsing. Do the reverse for "backward" browsing. Store the key of the first record for future "forward" browsing and each key after that for future "backward" browsing.

PROCESS OPTIONS AND STATEMENTS USED FOR BROWSING

CSP provides one special statement and several Process Options for use when browsing. These statements set up the sequential processing, get the next record(s), and end the sequential processing. The basic cycle of events is begun by a CSP statement. I don't know why this operation is done with a statement and not a Process Option. The designers must have their reasons. Figure 11.5 illustrates the CSP statement used to begin CSP's sequential processing of an INDEXED file or DL/I segment with a specific key.

Move a starting value into the record's key field before issuing the "SET recordnam SCAN" statement. This statement does not retrieve data, it just points to the first record that will be read. In fact, this statement does not even change the record's return code! (I'd apologize, but *I* didn't design it!) I suggest that you create an EXECUTE process consisting of very little other than the two lines illustrated in Figure 11.5. This allows you to treat this function as a module and helps to clarify your logic.

Once the initial record has been pointed to, you may begin reading the data sequentially. CSP provides two Process Options for reading data sequentially, "SCAN" (forward) and "SCANBACK" (backward). SCAN may be used with all record

---

SET recordname SCAN, establish pointer for SCAN statement
- move a starting value into the key field before issuing command
- does not retrieve data
- only valid for indexed or DL/I segment records

CAUTION: This is a STATEMENT, not a PROCESS option, it is normally coded inside an EXECUTE process

```
..............................STATEMENT DEFINITION..............................
001   MOVE mapnam.field TO recdnam.keyfield;       load key for I/O
002   SET recdnam SCAN;
```

Figure 11.5.   Using SET to position file.

APPLICATION PROCESS LIST			
SEL PROCESS	OPTION	OBJECT	ERROR
000 name_of_routine	SCAN	record_name	EZERTN

Figure 11.6.   SCAN process option.

types, including database records. SCANBACK is limited to IN-DEXED records only.

Look at Figure 11.6. Since SCAN is an I-O process option, both an OBJECT and ERROR routine should be specified for the process. SCAN reads the next sequential record into the record description for processing by the application. Which record is "next" depends on what preceded the SCAN. SCAN gets first record in file/database if no previous I/O took place. If INQUIRY or UPDATE were executed previously, SCAN begins reading with the record following the record processed by the INQUIRY or UPDATE. If an ADD was executed previously, the record following the record just added is returned by SCAN. Following a successful SET record SCAN, the record is pointed to by the SET. After a successful SETINQ or SETUPD, SCAN gets the first row satisfying the process's select statement appears (SQL records only). After a previous successful SCAN, the next SCAN returns the record following the last record retrieved (under CICS this may be impacted by SEGMENTATION, a topic introduced later in this book). SCAN sets the file response code so that normal testing can be done to gauge the success or failure of the request. SCAN returns EOF at the end of data (NRF for relative files). Like most programming tools, EOF occurs when you attempt to read (SCAN) one more time after processing the last record.

When processing INDEXED records, the SCANBACK process option is available (Figure 11.7). This allows backwards reading of the file. Since SCANBACK is an I-O process option, both an OBJECT and ERROR routine should be specified for the process. SCANBACK reads the previous record from file. Because backward reading is harder for the system than forward reading

APPLICATION PROCESS LIST			
SEL PROCESS	OPTION	OBJECT	ERROR
000 name_of_routine	SCANBACK	record_name	EZERTN

Figure 11.7.   SCANBACK process option.

(don't ask—it doesn't make sense, but it is true), CSP (actually VSAM) requires that the previous I/O or SET operation used a key value that exists on the file. If the previous I/O or SET key did not exist, the NRF condition is raised. Like SCAN, SCANBACK defines "previous" record depending upon what preceded its execution. If an INQUIRY or UPDATE were executed previously, SCANBACK begins reading with the record in front of the record processed by the INQUIRY or UPDATE. If an ADD was executed previously, the record before the record just added is returned by SCANBACK. Following a previous successful SCANBACK, it gets the record in front of the record just read. Following a SET record SCAN, SCANBACK gets the record pointed to by the SET. Don't forget to check the return code; SCANBACK returns EOF at the end of data (yes, at the beginning of the file!).

If a SCAN or SCANBACK returns the EOF status (or NRF), CSP closes the sequential operation automatically. If the SCAN or SCANBACK are halted for logical reasons, you should CLOSE the sequential operations yourself. CSP provides the CLOSE process option for this purpose (Figure 11.8).

APPLICATION PROCESS LIST			
SEL PROCESS	OPTION	OBJECT	ERROR
000 name_of_routine	CLOSE	record_name	EZERTN

Figure 11.8. CLOSE process option.

CLOSE is an I-O process option; both an OBJECT and ERROR routine should be specified. CLOSE terminates a SCAN/SCANBACK process before the end of the file is reached. This frees unprocessed rows in a set of SQL rows retrieved, closes a file, or disconnects a printer. Don't forget that CLOSE is implicitly executed by CSP when a SCAN/SCANBACK loop reaches NRF or EOF. Executing it again will raise an error condition. CLOSE is also automatic when another I-O process is executed for the same object and when applications transfer to one another. Database commit or rollback operations automatically execute the CLOSE, too.

Two process options are designed specifically for SQL row records. SETINQ (Figure 11.9) begins sequential read-only processing, and SETUPD begins sequential reads combined with

APPLICATION PROCESS LIST			
SEL PROCESS	OPTION	OBJECT	ERROR
000 name_of_routine	SETINQ	record_name	EZERTN

Figure 11.9. SETINQ process option.

DELETE and REPLACE operations. Specify both an OBJECT (SQL row) and ERROR routine for SETINQ. SETINQ selects a set of rows from a database to be fetched later using SCAN (read-only capability). Since SQL data is being processed, SETINQ opens a CURSOR used by CSP. SETINQ points to the first row with a key greater than or equal to the key at the time of the SETINQ. SETINQ does not bring any data into the record for processing. Rows in the cursor are sorted by default into key sequence, unless the SQL statement has been modified (see Chapter 12 for SQL processing).

SETUPD (Figure 11.10) is basically the same as SETINQ but allows DELETE or REPLACE operations against rows (causes locking). SETUPD positions the result table to the first record with a key greater than or equal to the key at the time of the SETUPD. SETUPD does not bring any data into the record for processing. The output rows are NOT sorted by SETUPD (a SQL limitation).

APPLICATION PROCESS LIST			
SEL PROCESS	OPTION	OBJECT	ERROR
000 name_of_routine	SETUPD	record_name	EZERTN

Figure 11.10. SETUPD process option.

SEQUENTIAL PROCESSING SUMMARY

To perform "browsing" operations, do the following:

- Move the desired starting key to the key field. If accessing an INDEXED record, use "SET recname SCAN" inside an EXECUTE process. Otherwise, if accessing an SQL row record, use a SETINQ process.

  Both SET recname SCAN and SETINQ are used to position a pointer to the first record that will be returned by the next SCAN process or SCANBACK process (indexed files only). No record data is available to the application until after a SCAN or SCANBACK (indexed files only) is executed.

- To go forward: SCAN until the screen is full (you must count), or until NRF or EOF.

- To go backward: SCAN once (makes sure key is valid), then SCANBACK until screen is full (you must count again), or until EOF.
- CLOSE unless CSP has automatically closed already.

Don't forget to CONVERSE after filling the map with data! More than one application developer has left out that little detail. Figure 11.11 illustrates several different combinations of key values

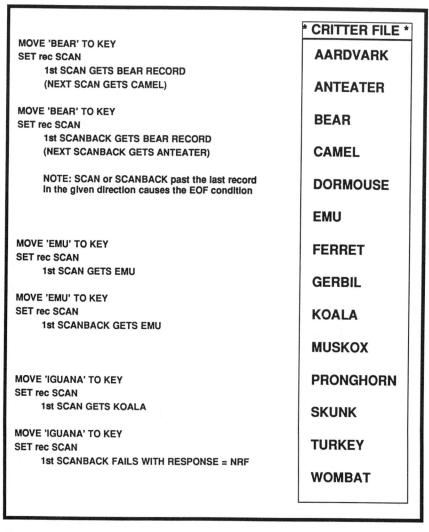

Figure 11.11. File positioning examples.

and CSP processing. Please note that in each case the illustration shows what would happen if the very *first* action after the "SET rec SCAN" is a SCAN or SCANBACK.

SCROLLING

This is a list of some possible methods for achieving parts of the scroll logic. There are many other possibilities.

1. Obtain initial key by:
   prompting the user via previous display, or retrieving it from WORKING STORAGE, or it might be passed from CALLing application.
2. Obtain browse direction by:
   getting the direction from a user prompt, or using function key tests, or default based upon design.
3. Set pointer by:
   SET record SCAN (indexed files), if processing an indexed file backwards. It may also be necessary to issue a SCAN to guarantee that record being pointed to is valid, or SETINQ (SQL database) OPENs CURSOR, does not fetch a row.
   It is possible to use other I/Os to initialize a pointer, but SET record SCAN and SETINQ are an obvious prelude to sequential processing (may make it easier for a maintenance programmer to comprehend your intent some time later).
4. Fetch rows one at a time using SCAN or SCANBACK for indexed files, depending upon the direction of the browse. For SQL records, the SELECT statement associated with the SETINQ may be pointed in the appropriate direction by a combination of the WHERE clause and the ORDER BY clause. For instance:
   (WHERE keycol <=:topkey . . . ORDER BY keycol DESC). SQL processing might require that a work file (or array in Working Storage or a CSP Data Table) be used to store data for performance purposes (in extreme cases).
5. Process each record's data. Bumping a counter as each SCAN or SCANBACK is executed provides a built-in subscript value for moving data to map fields.
6. End the process:
   CLOSE causes some system resources tied up by the sequential processing to be freed. Use EZECOMIT or EZEROLLB to

commit changes or rollback changes prior to freeing locks. CALL COMMIT or CALL RESET perform the same functions as EZECOMIT or EZEROLLB respectively when under CICS. CLOSE and CALL COMMIT are implicit when a MAIN application ends (EZECLOS, DXFR, XFER), at the end of an application segment in SEGMENTED mode (when a CONVERSE is executed), or when a DL/I PSB is released.

7. Again, don't forget to CONVERSE!

Figure 11.12 shows the process a little more graphically.

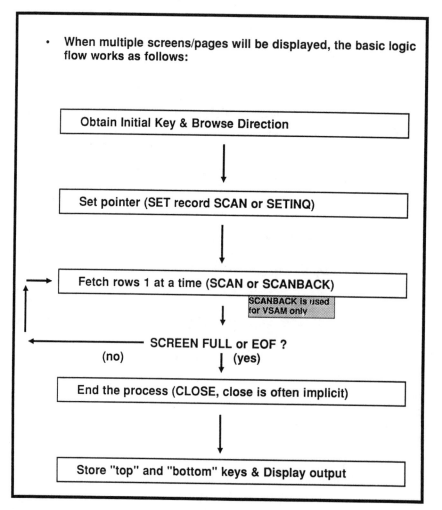

Figure 11.12. Multi-record panel processing.

CHAPTER 11 EXERCISES

Questions

1. What CSP process options are used to read data sequentially going forward or backward (respectively)?
2. What CSP statement is used to begin the sequential processing of an INDEXED record or DL/I segment?
3. What process options are used to begin the sequential processing of SQL row records?
4. When should a CLOSE be executed?
5. When should a CLOSE not be executed?
6. Name at least two things that will automatically CLOSE sequential processing.
7. What condition is raised when all records have been read (in either direction) during sequential processing?
8. What condition is raised if SCANBACK is issued and the previous I-O did not use a key that exists in the file?
9. What types of records may SCANBACK be used with?
10. What is a simple method to make sure that SCANBACK has a valid key to start with?

Computer Exercise
Browse of a VSAM file

In this exercise, you will modify your application to use the first screen displayed as before or to allow the user to fill a screen with multiple records.

1. Create a new map called BB01M02 (use the same map group you used previously). The new map should have a title, column headings, data from at least four records, and a message line. Include a text constant telling the user to press ENTER to return to the previous display (all fields except QUIT field should be ASKIP).
2. Modify your first map so that it includes four sets of directions (text constants) for the user:

   Press ENTER to display single record
   Press PF7 to scroll backwards (not activated yet)
   Press PF8 to scroll forwards
   Enter QUIT and press any key to exit

```
                             *** your title ***


        INVENTORY ID                    TITLE                  ON HAND
     _____        _____   _____

     XXXXXXXXXXXXXXXXXXXX    XXXXXXXXXXXXXXXXXXXXXXXXXXXXXXXXXXXXXXX   9,999

     XXXXXXXXXXXXXXXXXXXX    XXXXXXXXXXXXXXXXXXXXXXXXXXXXXXXXXXXXXXX   9,999

     XXXXXXXXXXXXXXXXXXXX    XXXXXXXXXXXXXXXXXXXXXXXXXXXXXXXXXXXXXXX   9,999

     XXXXXXXXXXXXXXXXXXXX    XXXXXXXXXXXXXXXXXXXXXXXXXXXXXXXXXXXXXXX   9,999


             Press ENTER to display single record
             Press PF7 to scroll backwards (not activated yet)
             Press PF8 to scroll forwards
             Enter QUIT and press any key to exit  =>  _____


  Enter = Return      PF3 = Quit      PF7 = Backward      PF8 = Forward
```

Figure 11.13.   Screen format specification.

3.   If the user presses ENTER, use the same logic as before.
4.   If the user presses PF7 or PF8, use the Inventory ID entered
     on the first screen to begin the browsing process. Fill the
     screen going FORWARD through the file (if you reach end of
     file, simply stop). Store the first Inventory ID displayed and
     the last Inventory ID displayed in WORKING STORAGE.
     Cause your application to enter the first time logic upon
     return from the multiple record display screen.
     **OPTIONAL PORTIONS OF THE ASSIGNMENT, DO
     THEM ONE AT A TIME ! ** (Some people find it easier to do
     item 6 first. Do what feels comfortable to you.)
5a. Make the browse work BACKWARDs (change the screen to
     indicate functionality); alter the PF7 logic for this.
     *Do the forward processing first and make sure that it works.
     Only then should you begin the backward processing.*
5b. Add text constants to the second map (multiple rec display)
     telling the user to press PF7 to go forward, PF8 to go back-
     ward, or ENTER to return to the Inventory ID entry map.

Possible pseudocode for exercise:

> **Define the map with fields subscripted accordingly (no, there is no shortcut to naming the fields)**
>
> **Define the record**
>
> **Move prompt asking for starting key to map**
>
> **CONVERSE map**
>
> **Use key entered to SET SCAN (KSDS) or SETINQ (SQL)**
>
> **WHILE not-eof and screen-has-room**
>
> > **PERFORM scan routine**
> > **move data to subscripted fields in map**
> > **bump subscript**
>
> **CLOSE if not-eof**
>
> **CONVERSE map with many records**

Figure 11.14. Sample pseudocode.

6a. Modify your application to browse forward or backward (using Inventory IDs stored in WORKING STORAGE) at the user's direction. If they press any function key other than PF7 or PF8, return them to the FIRST iteration of the application (redisplay the first map asking for a Inventory ID).

6b. Modify the application to return to the first (BB01M01) map if the user presses ENTER (to really get fancy, display the record matching the cursor position when they press ENTER).

12

SQL Record Processing

In recent years, one database model has become increasingly important, the Relational Model. The relational databases in the IBM world use Structured Query Language (SQL) for data definition, data manipulation, and data control. IBM's SQL databases include DB2 (MVS), SQL/DS (VM and VSE), SQL/400 (OS/400), and Data Manager (OS/2). CSP supports access to SQL databases in all environments that have them. This chapter addresses the use of SQL from a CSP application. If your installation does not use an SQL database, consider skipping this chapter. This chapter does not cover SQL database design or SQL coding. (Time to find another book!) This chapter *does* include what is required to make CSP work with an SQL database. Specific examples in this chapter deal with DB2.

CSP provides different interfaces to SQL databases, including the following:

1. SQL row records defining the results of a query
2. DB2-oriented PROCESS OPTIONS with built-in SQL that use the SQL row records (above)
3. SQLEXEC for "do it yourself" SQL

SQL ROW RECORD DEFINITION

SQL row records represent the RESULT of an SQL query. Records may represent queries involving more than one table or view (a "join"). You may only update, delete, or insert data for rows that are based on a single-base table. You may only insert (ADD) rows if all NOT NULL columns are included in the CSP record's definition. Many installations use SQL SYNONYMs rather than real table names in their statements. Synonyms are alternate names used by application developers to make referencing SQL tables simpler. (Be careful! The SYNONYMs defined by the BINDer are the SYNONYMs used by the Application Plan.) SQL row record definition uses the same basic series of panels as any other record type (Chapter 6).

See Figure 12.1. Name the table(s) to be used to create the rows processed by CSP. SQL tables typically have a two-node name, CREATOR followed by a period followed by the TABLE

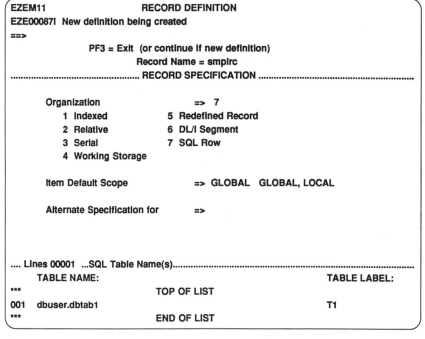

Figure 12.1. SQL row record specification.

NAME. When an SQL object such as a table is created, a creator ID (usually a user signon ID) is stored in SQL's catalog along with the table name. This way, two users might have different versions of the same table, each under their own user (creator) ID. If only a single name is given, SQL assumes that it is being given a table or view name that belongs to the current user ID or that it is being given a synonym name. Be sure you know the full table name (creatorID.tablename) before you start defining the SQL row record. Be careful that the CREATOR name you specify is the name that will be used at execution time. When using SYNONYM names, the userid that BINDs the application plan (DB2) is the userid whose SYNONYM value will be used.

CSP issues a short TABLE LABEL that may be used as a qualifier to reference SQL row data (T1.colval). You may modify this if you like (a good idea).

Item default scope (GLOBAL or LOCAL) means the same thing here that it means for other CSP records. GLOBAL data items are represented by members in an MSL and may be re-used. LOCAL data items are known only in the context of the record in which they are defined. Data item definition for SQL row records is also very similar to data item definition for other record types with a few notable exceptions. First, if the table(s) that you named on the previous panel is known to the catalog, CSP will automatically load the SQL ROW DEFINITION DATA ITEM DEFINITION panel with data items based upon the table(s) definition. Second, you need to tell CSP which columns are read-only and which columns are part of the key. Finally, an SQL CODE entry (Figure 12.2) is provided for defining SQL data that is of a type not normally known to CSP (like Floating Point).

After you name the table(s) to be used, CSP provides a list of ALL columns in that table. Be sure to delete all columns that will not be needed by your application. It is important to list *only* those columns needed by the application. To overdo the column selection is wasteful (causes extra database I-O and locking). Use line Move and Copy commands to reorganize the columns into the order you want. You may reorder the columns in any way you wish.

You may rename the CSP data items as you see fit, but do not alter the SQL COLUMN NAME listed on the second DATA ITEM DEFINITION panel. To see if your data item definitions

EZEM15		SQL ROW DEFINITION				More: ->

==>

		PF3 = Exit	PF4 = SQL Compare			

Record Name = dbrow   Default Scope = GLOBAL

Total lines 0004 .... DATA ITEM DEFINITION

....

	NAME	TYPE	LENGTH	DEC	BYTES	READ ONLY	KEY	SCOPE
***		TOP OF LIST						
001	COL1	CHA	00010		00010	YES	YES	LOCAL
002	COL4	CHA	00003		00003	NO	NO	LOCAL
003	COL12	PACK	00003		00002	NO	NO	LOCAL
004	COL3	BIN	00004		00002	NO	NO	LOCAL
***		END OF LIST						

Figure 12.2.   SQL row record data item definition, 1.

match the data types in the SQL catalog, or to see if any were left out, press PF4. When creating an SQL Row Record, the PF4 key is used to compare columns in use to DB2 tables and to allow browsing of other column names for selection purposes.

Only data items to be updated should be marked READ ONLY NO; all other columns should be marked READ ONLY YES. Primary key columns should normally be marked READ ONLY YES. Minimizing the number of READ ONLY NO columns will reduce system load and potential lock conditions. Updating a key column may cause SQL to *avoid* using an index based upon that column. Some installations define TWO CSP records when a key column is sometimes alterable, one for normal use with the key column(s) marked READ ONLY YES and another record specifically for updating/insertion that allows the key column to be modified.

Good SQL database design usually includes a process called "normalization" (sorry, different book). Part of this process entails identifying different record types (tables) and the keys that will be used to access them. The design process also identifies keys that are combinations of one or more data items (columns). For example, a personnel system might use three record types:

Employee
Employee Dependent
Employee/Dependent Health Claims

The key for Employee would probably be some kind of Employee ID (Social Security Number, for example). Each Employee Dependent might be identified by *combining* the Employee ID with some way to uniquely identify the dependent, maybe a Dependent Number.

Employee/Dependent Health Claims might be identified by:

Employee ID
Dependent Number (blank for actual employee)
Claim Date
Claim Number

These different portions of "key" data may be stored in noncontiguous fields!

Mark key column(s) as KEY YES, and CSP will use them in the default SQL statements generated by processes using this object. This is a very useful feature, and you should use it when you can.

At the top of the panel (upper right corner), a "More: –>" arrow shows that you may shift right (PF11/PF23) to see another page. This takes you to the second SQL ROW DEFINITION DATA ITEM DEFINITION panel (Figure 12.3).

This column lists the CSP data item name, the SQL column name, and a column called SQL CODE. The SQL CODEs describe data types for the columns. Except for FLOAT, SQL data types have corresponding values in CSP. LONG and FLOAT may be defined using the CSP HEX data type with a length of 8 (also requires SQL DATA CODE). HEX data items representing FLOAT columns in SQL should specify a SQL DATA CODE of "481" instead of the default value. This is the only time you should modify one of these codes. Very few installations are using FLOAT data types in their SQL databases, so this may not come up in your shop at all.

Notice that the upper right-hand corner of the panel now

```
EZEM15                      SQL ROW DEFINITION                    More: <->

==>
                   PF3 = Exit        PF4 = SQL Compare
           Record Name = dbrow                          Default Scope = GLOBAL
   Total lines 0004 ............................ DATA ITEM DEFINITION ..............................................
           NAME                        SQL COLUMN NAME                      SQL
                                                                           CODE

   ***                      TOP OF LIST
   001   COL1                          COL_ONE                             453
   002   COL4                          COL_FOUR                            453
   003   COL12                         COL_TWELVE                          485
   004   COL3                          COL_THREE                           501
   ***                      END OF LIST
```

Figure 12.3. SQL row record data item definition, 2.

shows a "More: <->" arrow pointing in both directions. This means that you may use PF10/PF22 (left) to shift left and see the first screen again or that you may press PF11/PF23 (right) to see the last data item definition panel.

The final SQL ROW DEFINITION DATA ITEM DEFINITION panel is used to add descriptions (documentation) to the CSP data items. While not necessary, it seems like a good idea, especially if your shop has naming standards that need to be explained to newcomers.

Once the data items have been defined, check the RECORD DEFINITION DEFAULT SELECTION CONDITIONS panel (Figure 12.4). This panel shows the default SELECT statement that will be used for all input operations using the SQL row record. It will also be the basis for the SQL statements created by I-O processes that modify the database.

On this panel you may alter the default statement's WHERE clause. This is crucial when joins are being performed to make sure that all proper conditions have been added. It is also wise to add WHERE clause items to make the SELECT as restrictive as possible.

When you use an SQL row record in a process, you may alter the SQL created to do the I-O. When viewing processes in the

```
 EZEM16                    SQL ROW RECORD DEFINITION
 EZ005901 You may edit lines preceded by line numbers
 ==>
   PF3 = File and exit  PF4 = Reset to default statement   PF5 = SQL syntax check
   Record  = smplrow
                                       Modified clause        = NO
 Total lines 0009 .......... DEFAULT SELECTION CONDITIONS DEFINITION ...........................
                         TOP OF LIST
 ***  SELECT
 ***     COL1, COL2, COL3
 ***  INTO
 ***     :COL1, :COL2, :COL3
 ***  FROM
 ***     CREATOR.TABLEA T1,
 ***     CREATOR.TABLEB T2
 ***  WHERE
 009 ;** INSERT DEFAULT SELECT CONDITIONS HERE **
 ***                         END OF LIST
```

Figure 12.4.   SQL row record default SQL statement.

APPLICATION PROCESS LIST or STRUCTURE LIST, place an "O" in the SEL column to tell CSP you wish to view/modify the SQL (Figure 12.5). The statement displayed on this panel is based upon the RECORD DEFINITION DEFAULT SELECTION CONDITIONS panel. To reset to the default, press PF4. You may alter (if you feel comfortable with SQL) the statement fairly radically if you choose to. The panel will list line numbers next to the lines that may be altered or deleted. If you change a statement and want CSP to do a syntax check, press PF5 (but be prepared with a manual listing SQL codes).

Two fields were added to this panel by CSP Version 3 Release 3: Execution Time Statement Build and Single Row SELECT. Execution Time Build may be set to YES or NO (NO is the default). Setting Execution Time Build to YES causes CSP to build the SQL using an EXECUTE IMMEDIATE. EXECUTE IMMEDIATE is a DYNAMIC SQL statement and must be parsed, syntax checked, and optimized each time it is executed. Most DBAs

```
EZEM3L                         APPLICATION DEFINITION
EZ005900I You may edit lines preceded by line numbers
==>
   PF3 = File and exit  PF4 = Reset to default statement  PF5 = SQL syntax check
   Process = sqlget                      Description = get sql row
   Option  = INQUIRY        Object = smplrow         Modified statements = NO
Total lines 0012 .......... OBJECT SELECTION:   SQL STATEMENT DEFINITION ........................


                 Execution Time Statement Build  => NO
                 Single Row SELECT               => NO

***                         TOP OF LIST
*** SELECT
002   T1.COL1, T1.COL2, T2COL3
***     INTO
004    :COL1, :COL2, :COL3
*** FROM
***    CREATOR.TABLEA T1,
***    CREATOR.TABLEB T2
*** WHERE
009   ( T1.COL1 = T2.COL1)
010   AND T1.COL2 > +93
011   AND T2.COL3 LIKE 'TR%'
***                         END OF LIST
```

Figure 12.5.   SQL row—Object selection panel.

frown on (to put it mildly) using DYNAMIC SQL without having received their prior blessing. DYNAMIC SQL is not necessarily bad, but in the wrong instances it can truly cause some performance problems.

Single-Row Select may also be set to YES or NO (again, NO is the default). Internally, CSP uses CURSORs (more on this later) for inquiries. CURSORs may not be as efficient as a single-row selection if only one row is desired. CSP Version 3 Release 3 allows the application developer to tell CSP that a query is expected to retrieve only one row from the database. This will enhance performance. *Caution!* If you mark Single-Row Select and the query somehow returns more than one row, your query will FAIL at execution time (bringing back NO data). Be sure that you make all decisions based upon your design.

SQLEXEC—"Roll your own SQL"

In addition to the CSP-generated SQL created as part of the normal process options, an additional process option is available for SQL users. SQLEXEC allows an SQL-savvy individual to create statements designed to improve performance or make maintenance easier. Common uses for SQLEXEC are mass UPDATE and DELETE statements based upon common criteria. Another good reason to use SQLEXEC is when an INSERT to the database will be performed based upon querying another table. Key in the embedded SQL statement using host variables where necessary. Do *not* define indicator variables. Remember that CSP provides the IF...NULL and SET...NULL for that. SQLEXEC is a process option just like INQUIRY, ADD, SETINQ, and all the others. The process statements will be split around the I-O activity created by your statement. The APPLICATION DEFINITION OBJECT SELECTION panel is used to key in the SQL statement.

SQLEXEC may be used to execute:

INSERT
DELETE
UPDATE
CREATE (for tables, views, and synonyms)
DROP (for tables, views, and synonyms)
GRANT
REVOKE

SQLEXEC may NOT be used for:

SELECT (use INQUIRY, UPDATE, SETINQ, or SETUPD processes)
COMMIT (use EZECOMIT)
ROLLBACK (use EZEROLLB)
DECLARE CURSOR, OPEN, FETCH, or CLOSE
INCLUDE
PREPARE, EXECUTE, or EXECUTE IMMEDIATE
DESCRIBE
WHENEVER

A complete list of allowable modifications or deletions may be found in the CSP/AD Developing Applications manual.

Like other SQL process options, SQLEXEC also allows specification of Execution Time Statement Build. You should still get the DBA's permission to use this feature if it is needed (very seldom is this a necessary thing).

CSP Version 3 Release 3 added another new feature to SQLEXEC, the MODEL statement capability. By entering the type of statement you are interested in on the "Model SQL Statement Generation" line, CSP will build the basic syntax for you to start with. This is especially handy if you are not normally a big SQL user. Finally, remember that you can press PF5 at any time to have the SQL syntax checked. Be sure to use this feature often.

Some Examples of SQLEXEC

When using the SQLEXEC process option, both BEFORE or AFTER processing still apply. The SQL statement is coded on the Default Specifications Panel. If host variables are used only in SELECT lists, UPDATE . . . SET, INSERT VALUES, WHERE clauses, and HAVING clauses, then static SQL generation is possible (Figure 12.6).

If host variables are used instead of DB2 object names or SQL keywords, the SQLEXEC process uses dynamic SQL to PREPARE the statement, rendering static generation of the statement moot (Figure 12.7). Use SQLEXEC to perform inserts not easily performed using ADD. Whenever multiple updates or de-

```
UPDATE  MY.TABLE
        SET  PLATE  =  :CHINATYPE
        WHERE  GUEST  =  :BOSS.ORNOT
```

Figure 12.6. Sample STATIC SQL statement.

```
UPDATE  :uinput.tablename
        SET  :uinput.colnam  =  100
        WHERE  :uinput.colnam  <  100
```

Figure 12.7. Sample DYNAMIC SQL statement.

letes can be performed with a single SQL call, use SQLEXEC. Use SQLEXEC to directly update rows to minimize deadlocks resulting from cursor processing. Finally, use SQLEXEC to perform SQL data definition (CREATE, DROP, and so on).

A REVIEW OF SQL ROW RECORD DEFINITION HIGHLIGHTS

SQL row records represent the *result* of a query. Records may represent queries involving multiple tables. Mark as many columns as possible READ ONLY YES. Primary key fields should not be marked READ ONLY NO, unless a special, controlled record name is set up just to allow manipulation of that field. Minimize READ ONLY NO columns to avoid locks. Be sure to specify conditions for all joins. Try to avoid cursors that cause creation of temporary table (a SORT shows up in the EXPLAIN output).

Under Version 3 Release 3 and later releases, use the default SQL specification to limit a single-row select and avoid a cursor. Use PF11/PF23 to view/change the SQL DATA CODE and SQL COLUMN NAME (note change in More arrow(s) as you shift left/ right). Use PF11/PF23 again to view/change the field DESCRIPTIONs. Use PF10/PF22 to shift left. Leave SQL DATA CODE alone unless processing includes FLOAT data. In that case, call it HEX with length 8 and alter the SQL DATA CODE to 481.

SQL Locks

Most CSP processing of SQL row records is done via cursors (unless SINGLE ROW SELECT is specified). When a cursor is opened (by SCAN, SETINQ, SETUPD) rows are locked based upon the Isolation Level specified:

- Cursor Stability (the default since V3.2) allows maximum concurrent use by locking only those rows on the same page as the current row in the cursor.
- Repeatable Read reduces concurrency, because ALL rows read so far are locked until a SYNCPOINT is reached.

The database will prevent two users from modifying the same record. For instance, if User-1 has modified a row but has not yet committed/rolled back, User-2 will have read access to the OLD version of the row. If User-2 attempts to modify the row, the system will force them to wait.

Cursors

SQL's ability to handle SETS of data throws a "monkey wrench" into the processing of merely mortal traditional data processing methods. Most languages (including those at the innards of CSP) are not capable of dealing with more than one record at time. The makers of SQL (clever souls) created a logical construct called a CURSOR to take up the slack. CURSORs allow a lower-level tool to look at the result rows from a SELECT statement one row at a time. The CURSOR represents the SQL statement to be executed (not unlike a VIEW). The SQL statement's execution is controlled by the SELECT criteria within it.

- Some SELECT statements can be satisfied one result row at a time, without further database processing CURSORs. This procedure is easier on the system since only those rows currently being looked at take up space and locks.
- Other SELECT statements require a degree of preprocessing before the first result row may be presented. Good illustrations are those SELECT statements using DISTINCT or AVG. SELECT statements that require that the result table be completed before presenting a single result row are accomplished by building a temporary table in virtual memory. The CURSOR then processes the temporary table one row at a time. If temporary tables are large, they can cause a performance bottleneck. If they are small, no big deal.

The possibility of large temporary tables hogging response time calls for a discussion of some techniques used to get around the problem.

Places to Hold Cursor Data

CICS provides a mechanism called Temporary Storage for the storing of work data. CSP uses these areas heavily for internal processing and storing of segment data between iterations of the transaction. There are two basic types of Temporary Storage, Main (in virtual memory), and Auxiliary (on disk). Main Temporary Storage is very fast, but in a memory-short environment it is also very expensive. Auxiliary Temporary Storage is cheap but also disk-drive slow. Many XA and ESA shops have enough memory to use Main Temporary Storage heavily. Applications repeatedly causing creation of large temporary tables might consider calling a non-CSP application to process the SELECT statement and store all of the row record data at once for later processing (CSP does not provide a direct method for manipulation of Temporary Storage; that's coming in Version 4). CSP tables are also sometimes used to store cursor results, providing a quick and easy solution. CSP tables are limited to 4092 rows. Storing table data in this fashion assumes that up-to-date data is not needed by the users. CSP tables are temporary and will revert to their original form at the end of the application. One other place that may be used is an array in CSP working storage. This has the advantage of CICS Temporary Storage but without the extra call to an external module.

Any way you do it, a good idea is to read only a few pages' worth of data at a time. Few users scroll through more than four or five pages of data at a sitting. So pick an arbitrary number of pages and only read ahead that far. If the user exhausts the data, simply go get more.

Static vs. Dynamic CICS calls

SQL use can be separated into two groups of calls:

dynamic calls
static calls

Dynamic execution is the default, is used to test applications, and is used for infrequently run applications.

Before an SQL statement is executed, it must be parsed, checked for errors, and optimized; this process is called BINDing. Applications relying upon dynamic processing BIND each statement every time they execute (kind of expensive). Static calls assume that a BIND has been run once and the result stored as an Application Plan. Static calls allow execution multiple times without incurring additional BIND processing (not all environments support STATIC calls). This is similar to the difference between interpretive programming languages and compiled programming languages. Most of the time, static calls to SQL greatly outperform dynamic calls. It is possible to create a "static" load module for an application allowing better use of the database.

Dynamic execution is the default; it is used to test applications and for infrequently run applications. Because SQL statements must be validated and prepared each time the application executes, it is not the most efficient way to do normal processing. Static execution is set up as a part of the GENERATION process (see Chapter 14). Static modules are exported from CSP to a batch file. The batch file is then run through a fairly standard precompile, assembly, link-edit process culminating in a load module and an application plan. While it takes much more work, the efficiency gains usually make static execution worthwhile.

Three Other Important Things

Move 1 to the special variable EZEFEC to avoid abending when SQL errors occur. Otherwise, CSP sees a non-zero SQLCODE as a hardware error. (Again, I don't know why—that's just how it works.)

Instead of checking the SQL row record for NRF, EOF, and the other values, consider checking EZESQCOD. EZESQCOD contains the value of SQLCODE after each request (0=OK, 100=NRF/EOF, <0=ERROR).

If the application will be executed in SEGMENTED mode (see Chapter 18), move 1 to EZECNVCM to 1 before executing CONVERSE processes. This simulates CICS COMMIT/SYNCPOINT processing under TSO and during TEST. Without moving 1 to EZECNVCM the program will not work the same way in the test environment as it will in the execution environment.

UPDATING A RECORD

The next exercise asks you to update an SQL row record. The following list recaps the process necessary to perform updates, with a special tilt toward SQL processing.

1. Define map and record.
2. Clear map and record.
3. Move message to map requesting entry of record key (be sure to use EZEMSG field name).
4. Use CONVERSE to display initial panel and receive the user's input.
5. Use key entered to issue INQUIRY.
6. Move appropriate data to screen.
7. CONVERSE to display data for user update.
8. TEST . . . MODIFIED to decide if data was changed, and if necessary do additional edits.
9. If everything is okay, use UPDATE to get latest version of record.
   Test to make sure that record read by UPDATE matches record displayed after INQUIRY.
   Move modified data to record area.
   REPLACE record.
   CONVERSE with success message.

CSP-DB2 DATE MISMATCH

CSP and DB2 use different date formats. This makes it difficult sometimes to use DB2 dates with CSP applications. The following code may help. Figure 12.8 shows data item definitions that can be added to an application's working storage. Figure 12.9

CSP_DATE	10	NUM	00008	00008	(19yymmdd)
CSP_DATE_CCYY	15	CHA	00004	00004	(century+year)
CSP_DATE_MM	15	CHA	00002	00002	(month)
CSP_DATE_DD	15	CHA	00002	00002	(day)
SQL_DATE	10	CHA	00010	00010	(mm/dd/yyyy)
SQL_DATE_MM	15	CHA	00002	00002	(month)
SQL_DATE_SL1	15	CHA	00001	00001	
SQL_DATE_DD	15	CHA	00002	00002	(day)
SQL_DATE_SL2	15	CHA	00001	00001	
SQL_DATE_CCYY	15	CHA	00004	00004	(century+year)

Figure 12.8. Data items for SQL date conversion.

```
MOVE 0                      TO CSP_DATE    ; initialize field
MOVE MAPFIELD_DATE          TO CSP_DATE    ; date from map
CSP_DATE = CSP_DATE + 190000               ; get century
MOVE CSP_DATE_CCYY          TO SQL_DATE_CCYY; move fields
MOVE '/'                    TO SQL_DATE_SL1;
MOVE CSP_DATE_MM            TO SQL_DATE_MM;
MOVE '/'                    TO SQL_DATE_SL2;
MOVE CSP_DATE_DD            TO SQL_DATE_DD;
```

Figure 12.9.   CSP statements for SQL date conversion.

includes the CSP statements necessary to move a CSP date into DB2. The reverse process may be used to go from DB2 to a CSP formatted date. The CSP-DB2 date inconsistency is being addressed in Version 4 of CSP/AD.

After doing the moves in Figure 12.9, SQL_DATE may be used as a host variable in SQL statements referencing date columns.

CHAPTER 12 EXERCISES

Questions

1. An SQL table name is usually made up of two parts. Describe them.
2. What format is used to enter SQL table names during the record definition process?
3. What function key is used to verify SQL data items against the database's catalog?
4. Must CSP data item names be identical to the names of the underlying SQL columns?
5. What panel allows modification of the default SELECT statement created as part of SQL row definition?
6. What panel allows modification of the SQL statement generated to perform an SQL I-O process? How do you get to it?
7. What process option allows a developer to key in his or her own SQL statement?
8. What special EZE variable may be tested to check the success or failure of SQL access?

9. What is the difference between STATIC and DYNAMIC SQL?
10. How may an application specify that SQL errors are not treated as hardware failures?

Computer Exercise
Viewing, Updating an SQL row record

In this exercise, you will have an opportunity to define and use an application using an SQL table:

1. Create a new screen called (BB02G BB02M01) containing:

ISBN	10 characters, unprotected, first nine must be numeric, last character may be alphanumeric (0–9, A–Z)
TITLE	40 characters, unprotected
PUBLISHER	40 characters, unprotected
REPLACEMENT-COST	7 characters, numeric, 2 decimals
SALE-PRICE	7 characters, numeric, 2 decimals
QUANTITY-ON-ORDER	6 characters, numeric, 0 decimals
QUANTITY-ON-HAND	6 characters, numeric, 0 decimals
QUITFLD	4 characters, unprotected
EZEMSG	78 characters, auto-skip

2. Define an SQL row record called BB02R01 using the _____.BOOKINV table. Choose only the following columns:

ISBN	(read only; this is the key field)
PUBLISHER	
TITLE	
REPL_COST	
SALE_PRICE	
QTY_ON_HAND	
QTY_ON_ORDER	

3. Design (and then code) a CSP application called BB02A to perform the following:
   a. Prompt user for ISBN
   b. Get data from the appropriate database row (use IN-QUIRY process option)
   c. Display BOOKINV data (ignore any changes for now)
   d. Allow continued processing until PF3 is pressed or QUITFLD = "QUIT"

   **** OPTIONAL **** make steps a–d work first ****

   e. If your table has been set up to allow UPDATEs:
      If data was entered and is correct, get the latest row data using the UPDATE process option, move new data into the record, change the database using the REPLACE process option, and display success message.

      Otherwise, if the data is incorrect, display an error message and ask the user to try again.

13

Transferring Control

It is frequently desirable for a CSP application to decide to execute another CSP application or even a non-CSP program. This design technique allows reuse of application/programs that serve specific purposes. This design concept also allows a designer to isolate one application's logic from another, the so-called "black box" idea.

CSP allows an application to transfer control to another quickly and easily using the CALL, XFER, or DXFR commands. CSP may also transfer control to other, non-CSP programs. Transferring to non-CSP programs allows the use of existing calculations, manipulation logic, or fitting CSP applications into an existing system. It also allows a designer to call non-CSP programs to do functions that CSP cannot.

CSP allows control to be passed from application to application or even to non-CSP programs using the following variety of mechanisms:

CALL
XFER
DXFR
CALL CREATX (CICS only)

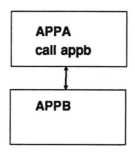

Figure 13.1. CALL—Logical path.

The logical functions of these different mechanisms remains the same from environment to environment. The exact functioning of the different methods varies based upon the CSP execution environment. In all cases, data may be passed as control is transferred. This chapter discusses the different control transfer capabilities and their merits.

CALL allows the execution of subroutine programs with execution returning to the CALLing program immediately upon completion of the CALLed program/application. Figure 13.1 illustrates the subordinate relationship of the CALLed routine. CALLs are used most frequently for logical subfunctions like calculations and data manipulations.

XFER and DXFR cause the current application to stop immediately and cause another application/program to begin. XFER and DXFR work identically in TSO, CMS, and OS/400. In CICS, DXFR to a NON-CSP program causes a CICS XCTL to execute, and XFER causes a CICS START to execute due to CICS-DB2 restrictions. DB2 programs that DXFR to one another are frequently part of the same Application Plan. This is sometimes avoided by using EZECOMIT and DB2 Dynamic Plan Selection. The advent of packages in DB2 Version 2 Release 3 will make it much simpler to have all DB2 applications that might transfer to one another part of the same application plan.

Segmented CICS applications may also DXFR and set EZESEGTR to a new CSP trans-ID value prior to issuing CONVERSE. DXFR is much easier on CICS than XFER and is preferred. DPPX systems perform an XFER by combining a DTMS

CREATX followed by a NXTRANS; a DXFR is more efficient. Figure 13.2 shows that applications/modules that XFER or DXFR to one another are at the same level execution-wise. This construct is used when transferring from one logical function to another, for instance, a menu program transfers to a sub-menu.

CALL, XFER, and DXFR are all CSP statements. They are coded inside of processes of statement groups. Part of the application's design process should be to determine when and how control should be transferred (if it is necessary to meet the application's objectives). Pay particular attention to CSP documentation for your environment on what the most effective and efficient transfer mechanism might be in a given situation.

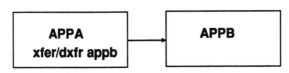

Figure 13.2.   XFER/DXFR—Logical path.

CALL

CSP allows an application to execute another application or a non-CSP program as a subroutine. Control temporarily transfers from the CALLing application to the CALLed program/application. When the CALLed program/application finishes, the CSP application resumes execution at the statement that follows the CALL (Figure 13.3).

```
CALL subrtn [arg1,arg2,...argn]
   (     NOMAPS
         NONCSP
         REPLY
```

Figure 13.3.   CALL statement syntax.

In the call, subrtn may be the name of a CSP application, an EZE subroutine (EZECOMIT, EZEROLLB, EZEC10, EZEC11, EZEG10, EZEG11), or a non-CSP program. EZECOMIT and

EZEROLLB calls are important, especially when processing DB2 data. EZEC10, EZEC11, EZEG10, and EZEG11 are used to check or generate modulus 10/11 values. The CSP documentation has complete information about the use of the EZE routines.

Arg1-argn may be passed to the calling program/application as parameters. CSP limits the passing of data in arguments, level-77 working storage data items, map names, record names, and working storage record names. It is often easiest to pass the name of a working storage record.

NOMAPS, NONCSP, and REPLY may be omitted or used in any combination. NOMAPS tells CSP that the subroutine will not display anything on the terminal. NONCSP tells CSP that the subroutine is not a CSP application. This speeds processing by causing CSP to skip the ALF (which contains CSP applications) while searching for the program and go directly to the system program libraries. REPLY makes it possible for the application to continue if the subroutine generates a non-zero return code (available in EZERT8). If REPLY is not specified and a non-zero return code is generated, then the CALLing application is terminated. CALL may not be used within a flow stage.

Again, CALL is branch-and-return processing. Control will return to the CALLing module immediately upon completion of the CALLed routine, at the statement immediately following the CALL.

XFER

XFER transfers control to another program immediately. The program/application currently being executed halts and the new program/application begins. Figure 13.4 illustrates the syntax for the XFER statement.

XFER tranid or cspappl or progname or EZEAPP
wsrecord
NONCSP

Figure 13.4.   XFER statement syntax.

The object to be transferred to varies based upon the execution environment. Non-CICS applications may XFER to CSP application names (cspappl) or non-CSP program names (progname). CICS applications and DPPX applications may XFER only to transaction codes (tranid, 1-4 chars. in CICS, 108 chars. in DPPX).

Regardless of the environment, the special CSP variable EZEAPP may be used as the transfer object. The name of the object to be transferred to is moved into EZEAPP before the XFER. EZEAPP allows a degree of portability for XFER logic.

Only one working storage record may be passed per XFER (wsrecord). CSP passes one address, containing all data described in that working storage record *except* 77-level data items. The extra working storage record must be defined to your CSP application using the TABLE AND ADDITIONAL RECORDS LIST panel.

XFERed-to applications must not be defined as a CALLed application. CALLed programs *may not* execute an XFER. In CICS, XFER causes a CICS "START" to be executed. This can be hard on CICS performance and is not normally recommended. Older DB2 applications might do this to get around an earlier problem with CICS SQL use (more on this later). DXFR and XFER work almost identically in all environments except CICS and DPPX.

Like CALL and DXFR, XFER allows use of the NONCSP option. NONCSP saves time because CSP looks immediately in the program libraries, rather than searching the ALF(s) for the object to be transferred to.

DXFR

DXFR transfers control to another program immediately—that is, the program/application currently being executed halts and the new program/application begins. Logically, DXFR and XFER are interchangeable. In CICS and DPPX systems, the impact of DXFR is far less than the impact of XFER, and DXFR should be used instead (Figure 13.5).

```
DXFR cspappl or progname or EZEAPP
              wsrecord
              NONCSP
```

Figure 13.5.   DXFR statement syntax.

The object to be transferred to by DXFR may be either a CSP application (cspappl) or a non-CSP program (progname). Like XFER, DXFR may use the special CSP variable EZEAPP to specify the transfer object's name. The name of the object to be transferred to is moved into EZEAPP before the DXFR. EZEAPP allows a degree of portability for DXFR logic.

Only one working storage record may be passed per DXFR (wsrecord). CSP passes one address, containing all data described in that working storage record *except* 77-level data items. You may also transfer control without passing data. The extra working storage record must be defined to your CSP application using the TABLE AND ADDITIONAL RECORDS LIST panel.

Like CALL and XFER, DXFR allows use of the NONCSP option. NONCSP saves time because CSP looks immediately in the program libraries instead of searching the ALF(s) for the object to be transferred to.

DXFRed-to applications must not be defined as a CALLed application. CALLed programs *may not* execute an DXFR. DXFR and XFER work almost identically except in the CICS and DPPX environments. In both CICS and DPPX, the DXFR provides a more efficient transfer mechanism.

CALL CREATX

When executing under the CICS environment, it is possible to cause the execution of an asynchronous task. CSP supports this by including the special command CALL CREATX (Figure 13.6).

```
CALL CREATX recdname,tdqq,termid
```

Figure 13.6.   CALL CREATX statement syntax.

CALL CREATX causes a CICS START command to be executed. This begins execution of another CICS transaction (CSP or non-CSP) that is completely separate from the current application (asynchronous). The command requires two parameters; the third is optional. The first parameter (recdname) names a working storage record that must have the following format:

- two-byte binary field, containing length of record (including this length field, max. size 4095 bytes)
- four-byte transaction code to be executed (character)
- four-byte blank field
- string passed as working storage to the initiated transaction

The second parameter (tdqq) names a four-byte level-77 data item describing the printer ID to be used. This data item contains one of two things: a four-byte identifier naming a CICS Transient Data queue (ask your CICS system programmer for information), or it may contain binary zeros if no printing is to be done.

The third parameter (termid) is optional, but, if used, it should also be a four-byte level-77 data item. If this parameter contains binary zeros or is omitted, CICS starts the new transaction without connecting it to any terminal. If this parameter contains a CICS terminal ID (again, ask the CICS system programmer), the new task will be associated with the designated terminal.

COMPARISON OF CALL/XFER/DXFR

Call

CALL is intended for use when the routine to be executed is a logical subordinate to the application currently executing. For instance, several applications might make use of a CALLed application to perform complex calculations. Another good reason to use CALL is when preexisting, non-CSP subroutines (calculations, edits, and so forth) are available. Finally, you need to CALL non-CSP routines when multidimensional tables must be processed or you need to do something that CSP cannot do.

XFER and DXFR

Use XFER and DXFR to transfer processing to a different function. Transfers are generally more efficient than CALLs if the second program will be short-lived or is large. In most systems, XFER and DXFR are equivalent, but you may only XFER to a tranID in CICS and DPPX. In DPPX systems, DXFR places much less load on the system than XFER. In CICS, the XCTL of a DXFR is generally easier on the system and response time than XFER's START and is preferred.

CICS allows only one application plan at a time for each transaction. Under Version 2 of DB2, dynamic plan selection allows each application to have its own application plan provided that a CICS SYNCPOINT occurs between application executions. It is common to combine the DBRMs from one or more Static SQL Modules together into a single plan when two applications use SQL DXFR back and forth (DB2 V2 R3 Packages will make this easier). It is also common to use EZESEGTR after a DXFR to a CICS CSP application and before a CONVERSE to cause processing to return to another transaction ID and plan (this often requires the use of parameter entry groups; see Chapter 18).

PASSING DATA

When CALLing another program/application, multiple addresses may be passed. Passed data is *not* copied but its address is used by the CALLed routine. When the CALLed routine uses and/or manipulates the data, it is actually using data belonging to the CALLing routine. XFER and DXFR allow only one data area to be passed. Since the transferring application will end, the passed data is copied and the new address is made available to the new application.

When using CALL, XFER, or DXFR to a non-CSP routine under CICS, only one address is passed. That data becomes available to the second program using the COMMAREA. The application being passed to (via CALL, XFER, or DXFR) must have a working storage record defined that matches the structure of the data being passed. This must be the working storage record identified on the APPLICATION DEFINITION APPLI-CATION SPECIFICATIONS panel.

CALLING NON-CSP ROUTINES

When calling non-CSP applications from CSP, you may pass data, too. The format of the data being passed is slightly different under CICS than under MVS/TSO, VM/CMS, or batch. When passing data with a CALL, XFER, or DXFR, the address of the data item or working storage record is passed to the non-CSP program. The non-CSP program must make sure the address is mapped correctly. Under MVS/TSO, VM/CMS, and batch, a four-byte (full-word) binary field is used to end the list of addresses. The four bytes must contain x'FFFFFFFF' (all hexadecimal 'f's). If the field is defined as a binary integer, the value will be −1 (minus 1).

CALLING CSP FROM NON-CSP

CSP may also be called by non-CSP routines. The CSP/AE module (usually DCBINIT) is the program executed, and it must receive two addresses. The first address is for a sixteen-byte data area holding an CSP ALF name in the first eight bytes and the CSP application name in the second eight bytes. The second address contains a structure to be passed to the CSP application as its working storage record.

SPECIAL CICS DB2 CONSIDERATIONS

The workings of DXFR in the CICS environment require a little extra explanation. If you will not be working in the CICS environment, skip this explanation. DXFR from a CSP application to a non-CSP application causes a CICS XCTL to be executed. This means that the CSP/AE program ends and the non-CSP program begins. DXFR is probably the best way to go from a CSP function to a non-CSP function. When transferring to another CSP application, no outward change is evident to CICS. CSP simply loads another module and branches internally. This can be a problem when running CSP under CICS to access DB2 data (alphabet soup!). As will be explained in Chapter 18, CICS allows a transaction to execute only one application plan at a time. Since most CICS-DB2 applications will be using "static" DB2 for performance reasons, each application's DB2 logic will be represented by something called a DBRM (Data Base Request Module). The DBRM is input to a process called the BIND where DB2 optimizes (as best it can) the SQL statements to be executed and stores the optimized logic into something called an application plan. The application plan must be available at the time of execution, or the DB2 portion will fail. CICS system programmers generate a special control table for each CICS "region" or address space called the RCT (Resource Control Table). The RCT tells CICS which transaction codes will use DB2. The transaction code entries in the RCT may point to one of two things, a DB2 application plan name or the name of a Plan Exit Program. If an application associated with an application plan name in the RCT uses DB2, DB2 loads the application plan named in the RCT. If an application associated with a Plan Exit Program uses DB2, the first time the application touches DB2 after a COMMIT (CICS SYNCPOINT), the plan exit program will execute and decide which application plan to call. If a DB2-using CSP/CICS transaction DXFRs to another DB2-using application or program, the system must be able to accommodate *both* sets of SQL statements. The application designer has two choices: large application plans or plan switching. The easiest choice is to have the DBRMs for all applications and programs that might DXFR to one another represented in a single application plan (DB2's bind procedure allows the specification of multiple DBRMs). In the case of using a large application plan, the RCT entries for all concerned transactions will point to the same application plan name. The other choice is to use a plan exit program (pointed to by each transaction's RCT entry) to decide which application plan to execute. This requires that the logic for applications be modified to include EZECOMIT executions to force the CICS SYNCPOINT needed to trigger the plan exit program the next time DB2 is touched. To avoid this problem entirely, some older systems use the XFER. XFER causes a CICS "START" ending the current CICS transaction and beginning a new one. The new transaction's RCT entry gets the appropriate plan. This is not a good solution CICS-wise.

Large Application Plans

Using the large application plan is by far the easiest method programmatically, and no code changes have to be made to support it. Unfortunately, if many DBRMs have to get processed and if many transaction IDs all point to the same application plan, this can cause long bind times and lockout problems. DB2 Version 2 Release 3 addresses this by allowing the use of "packages." In the new version of DB2, application plans may be made up of "packages" rather than DBRMs. When an application is altered, the DBRM is once again processed by the BIND, but the output is a "package" rather than an application plan. Another bind is done (usually only once) to pull all of the packages together into an application plan. When the bind changes packages in the future, the application plan uses them without the application plan being bound again. This explanation is fairly simplistic, but it gives you an idea of the complexity of the DB2/CICS matchup that CSP must live within.

Dynamic Plan Selection

Having a Plan Exit Program pointed to by the RCT allows the use of a DB2 feature called Dynamic Plan Selection. At the first use of DB2 after a CICS SYNCPOINT, the plan exit program is executed. The name of the DBRM used by the SQL statement is passed to the plan exit program and it then decides which application plan to ask DB2 to load. This method works well, but it does have two inherent problems. First, application logic must be modified to allow execution of EZECOMIT to force the CICS SYNCPOINT whenever an application plan must be "switched." Second, the act of switching plans causes a CICS-DB2 thread to be ended and another created. This is not nearly as bad performance-wise as issuing the START caused by XFER, but it is not as good as keeping the same application plan.

Another Way

CSP Version 3 Release 3 provides a third way to DXFR from one DB2 using transaction to another. The method requires that all DB2 access be done either *before* or *after* a map CONVERSE when using CICS SEGMENTED transactions. This forces the DB2 access into one transaction or another. Everything that happens before a CONVERSE is in one CICS transaction; everything that happens after the CONVERSE returns to CSP is in a second CICS transaction.

CSP provides a special variable named EZESEGTR. When segmented transactions (described in Chapter 18) reach a CONVERSE, one CICS transaction ends and another begins when the user responds to the map displayed. By changing the value in EZESEGTR prior to a CONVERSE, you can cause the transaction ID used for application plan selection to change. This has no effect on CSP since the CSP execution module is called regardless, but it does allow plan switching to occur at the time of each CONVERSE. To accomplish this successfully, DB2 I-O for an application must be done on only one "side" of a CONVERSE. This method is commonly used when the need to switch application plans arises in CICS-DB2 applications.

CHAPTER 13 EXERCISES

Questions

1. What commands are used to transfer control to other CSP applications?
2. Which CSP command is used to temporarily branch to a subordinate routine to do some work, then to return and finish the original routine?
3. May CSP use routines written in other languages?
4. Which CSP commands may be used to end the current application and begin executing another?
5. Where is passed data stored in the application being transferred from?
6. Where is the passed data stored in the application being transferred to?

7. XFER and DXFR work differently under CICS and DPPX. Which is generally preferred and why?
8. DXFR from one CICS CSP routine involving DB2 to another that involves DB2 sometimes adds complexity. What special CSP field is used to let the CICS-DB2 environment know that a change is taking place? When should the field's value be set?
9. What is the most common method used for easing the DXFR from one DB2 using routine to another?
10. What special command is used under CICS to execute another application asynchronously?

Computer Exercise
Transfer Using Menus

1. Create a menu application called BB03A. It should include a map (BB03G BB03M01) showing two choices to execute the two applications created in previous exercises.
2. Allow the menu user to press a function key or to enter a value to transfer to one of the previous applications.
3. Use DXFR to transfer control.
4. Test your application.
5. Create another application called BB04A that performs the edits currently being done in the first application. Remove the edits from the first application and replace them with a CALL to BB04A. This will require passing the Inventory ID entered and some kind of pass/fail indicator.
6. Test your results.

14

Application Generation and CSP/AD Utilities

Once you have developed and tested an application, it is time to produce it in executable form for the production environment. CSP/AD provides a function called GENERATE to create the executable application. The output from the GENERATION process is written to an Application Load File. If the ALF is unspecified, the system uses a default ALF called FZERSAM. The generation process creates several modules in the ALF. Once generation is complete, the application may be executed under CSP/AE. This is one of the most visible changes in CSP/AD Version 4.

Get into the generation process by entering "=Generate" on the command line of any screen, using option "G" from the LIST PROCESSOR, or choosing facility 4 (Generation) from the CSP/AD FACILITY SELECTION (main menu) panel (Figure 14.1).

This panel provides five choices, two that are used frequently.

1. Application Generation is the most frequently used selection. This generates modules for execution in the CSP/AE environment. Application generation includes the option to generate (or not) any maps or tables used by the application.
2. Map Group Generation is used to generate only a map group for use in the execution environment. This option is normally used only by CSP administrators.

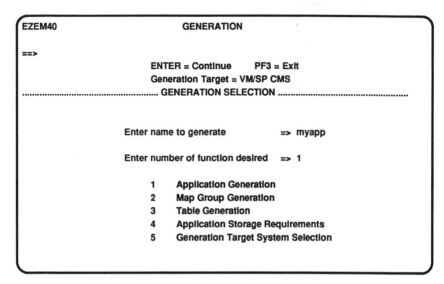

Figure 14.1. Generation selection.

3. Table Generation makes CSP data tables available in the execution environment. This is useful when modifying the contents of a shared table. This option is normally only used by CSP administrators.

4. Application Storage Requirements is a two-screen process used to see estimates of the number of bytes required by the various modules to be generated. This is certainly nice to know, but only your systems types will care.

5. Generation Target System Selection is the other frequently used choice. CSP allows you to choose which of the various potential target systems the generation is to generate code for. This allows you to generate on one system (say MVS/TSO) for execution in another (like CICS).

GENERATION TARGET SYSTEM SELECTION

It is necessary to choose which of CSP/AE's operating environments the application will be generated for. See Figure 14.2. CSP/AD automatically defaults the target system to match the system being used for generations. It is common to have CSP systems executed under different environments than they are

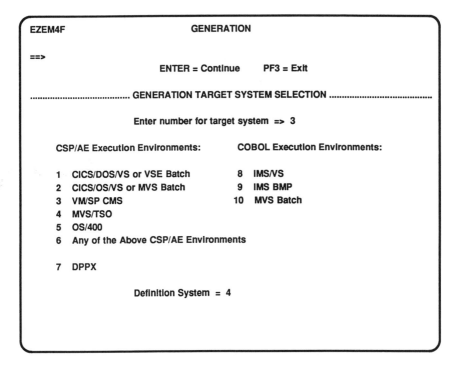

EZEM4F	GENERATION

==>

ENTER = Continue   PF3 = Exit

·· GENERATION TARGET SYSTEM SELECTION ··

Enter number for target system => 3

CSP/AE Execution Environments:	COBOL Execution Environments:
1 CICS/DOS/VS or VSE Batch	8 IMS/VS
2 CICS/OS/VS or MVS Batch	9 IMS BMP
3 VM/SP CMS	10 MVS Batch
4 MVS/TSO	
5 OS/400	
6 Any of the Above CSP/AE Environments	
7 DPPX	

Definition System = 4

Figure 14-2. Generation target system selection.

developed in. For instance, many installations run CSP/AD under TSO and execute the applications under CICS.

Enter the number of the target system selected. CSP will then
know to generate modules that are appropriate. Resist the temptation to mark option 6 (Any of the Above). This might seem
convenient, but the overhead is not worth it. Press ENTER to
continue.

Now you are ready to generate. But be careful! CSP puts the
cursor back on the application name on the GENERATION SE-
LECTION panel. Be sure to move the cursor before entering the
function number.

GENERATING THE APPLICATION

The panel presented at the beginning of the generation process is
the PREPROCESSOR OR GENERATION OPTIONS panel (Fig-

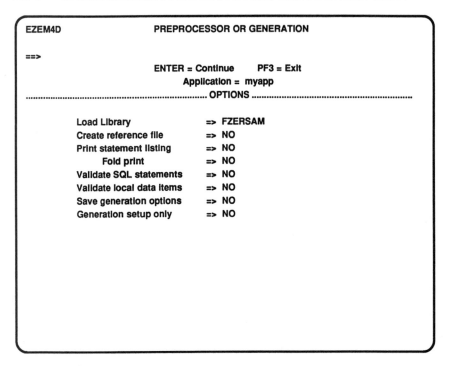

Figure 14.3. Generation options.

ure 14.3). Several specifics governing the generation of the application are controlled here.

First, enter the file name (ddname) of the ALF to be used to hold the generated application. If an ALF name is not specified, CSP defaults to FZERSAM. Either way, the file must already be allocated to the CSP session.

Create reference file (YES) causes generation to create a reference file named by suffixing the application name with an "R" (application TZ01 would generate TZ01$$$R; CSP expands the name to eight characters using the dollar sign "$"). If this is completed and the application later has an error during execution, the statement being executed at the time of the error is retrieved from the reference file and displayed.

Print statement listing (YES) causes printout to be generated. Fold print (YES) causes all data in the printout to be converted to upper-case letters. If turned on, the statement listing will also include any errors discovered during the generation process.

Validate SQL Statement (YES) causes CSP to invoke dynamic SQL to syntax check any SQL statements used in the application and makes sure that host variables used in the statements are compatible with the underlying SQL columns. Specify NO for SQL/DS since syntax checking is part of creating the access module.

Validate local data items (YES) checks any local data items against the current MSL list's data items. Error messages are issued if the definitions differ.

Save Generation Options (YES) causes the current set of generation options to be saved in the read-write MSL. This is useful when generating in batch later. Prior to CSP/AD Version 3 Release 3, saving once online was required for batch generation later. If the statement listing is turned on, the generation record appears as part of the output.

Generation Setup Only (YES) is used to store generation options in the MSL without actually generating. This is useful in environments where generation is done in batch for performance reasons by allowing the online developer to set the options to be used later.

Figure 14.4 shows two additional fields that appear when generating CICS applications, Execution Mode and Segmented Transaction. Choose the appropriate segmentation mode (segmented is usually desired for CICS transactions). Enter a segmented transaction ID; this is the transaction ID pointing to DCBRINIT in the CICS PCT (Program Control Table). No parameter group entries are required for this. *Caution:* If the application somehow moves zero (0) into the EZESEGM field prior to a CONVERSE, the application will run nonsegmented anyway. Chapter 18 contains detailed discussion about Segmented Transactions. Please refer there or to the CSP documentation for specifics.

ASSOCIATED OBJECTS

When generating an application, CSP will also ask if objects (tables, map groups, and so forth) are to be generated. Many of these objects may be shared with others, and if they have not

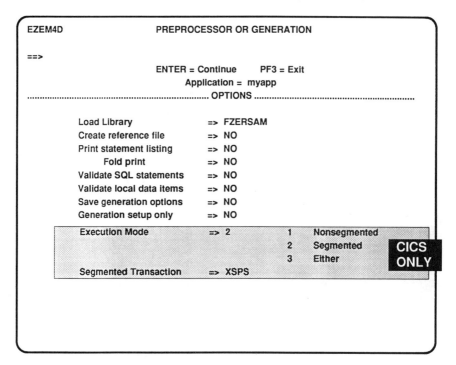

Figure 14.4. CICS generation options.

changed, they should not be regenerated. The panel that displays will be similar to Figure 14.5, but only choices pertinent to the application and the target environment will appear.

The TABLE option will be listed for each CSP table used by the application. If you are using an already generated table that has not changed, turn this off. If the table has been changed, or if it hasn't already been generated, two other options are important. To share tables under CICS, IMS/TM (IMS-DC), and DPPX, you must specify this option. While VM and TSO do not allow sharing, you must also mark SHARE yes if you want to make the table resident (more in Chapter 16). KEEP AFTER USE means that the contents of the table (as changed by the application) should not be deleted when the application ends. This option's use varies greatly from platform to platform and is further complicated by using the ALF utility (Chapter 16) to make a module resident. See Chapter 10 and the CSP documentation for more

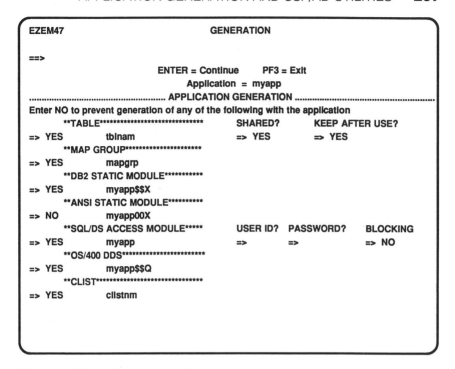

Figure 14.5. Generation of associated objects.

information on the valid combinations. The major usefulness of this option is when executing under CICS.

Each Map Group associated with the application will be represented by this panel. If the group already exists and has not been altered, then there is no point in generating again. If the group has not yet been generated, or if it has changed in *any* way, leave the default value of YES alone.

If SQL is being used, you may control what type of access module is created, and you may alter the name of the module to be created in the ALF. MVS users (CICS, MVS batch, TSO) must choose between DB2 STATIC and ANSI static modules. Most users of MVS systems choose DB2 STATIC MODULEs (creating Assembler source) over ANSI STATIC MODULEs (creating COBOL source). VM and VSE users must choose between SQL/DS ACCESS modules and ANSI STATIC modules; most users opt for the SQL/DS ACCESS MODULE. SQL/DS ACCESS MODULE users then have three other options:

1. To list the userID of the OWNER of the access module being created.
2. To specify the OWNER's password (this field is hidden with a dark attribute).
3. Whether to activate the BLOCK performance option in the generated code. BLOCK causes the application to process sequential table scans faster at the expense of more required memory. See your SQL/DS DBA for information on which alternative is best for you.

If OS/400 is the target environment, application generation may be used to generate the DDS (Data Description Specifications) for records used by the application. If specified, DDS will be created for all records that meet OS/400 conventions. See your system administrators for guidance.

The last item illustrated, CLIST, governs whether or not CLISTs should be generated for connecting and freeing files or printers under the DPPX system.

OUTPUT FROM GENERATION

The application generation process produces a variety of load modules. The generation output may be viewed using the ALF Utility described in Chapter 16. CSP pads most names out to a full seven characters, then adds a one-character suffix. Sometimes the fill character is a dollar sign ($) and sometimes it is a zero (0). Be careful! Application-naming conventions must account for the padding. Otherwise, application TZ01 and application TZ010 will result in duplicate output module names at generate time.

Each of the generation's possible outputs are listed below with a brief description of their contents:

ANSI SQL Static Module

This is a COBOL source module holding ANSI standard SQL statements representing the application's SQL. This SQL module must be PUT from the ALF and run through the standard

SQL preparation process for execution. If the application name is less than seven characters long, CSP will pad with zeros and add an "X" at the end to name this module (TZ01A becomes TZ01A00X).

Application Image Module (AIM)

This module contains application specifications and is loaded first when the application is executed. Its name is a duplicate of the application name.

Data Characteristics Load Module

This module contains the Data Item Characteristics table (DIT) that describes characteristics for all data items referenced in the application. Its name is padded to seven characters using dollar signs ($) and a suffix of "I" is added to the end (TZ01A becomes TZ01A$$I).

Data Description Specification (DDS) Load Module

Data needed to build DDS source members for OS/400 systems are listed in this module. Names are padded with zeros (0) and a suffix of "Q" is added (TZ01A becomes TZ01A00Q).

Data Load Module

The data load module contains tables for the Data Structure and Data Item Locations. They are copied into read/write memory during execution. This is where the records, maps, tables, and data items are defined. The name of this module is padded with dollar signs ($) and suffixed with a "D" (TZ01A becomes TZ01A$$D).

DB2 Static Module

The DB2 static load module is assembler source code for the static SQL statements in the application. The module must be PUT from the ALF and run through the standard DB2 source preparation routines for your system. The name of this module is padded with dollar signs ($) and suffixed with an "X" (TZ01A becomes TZ01A$$$X).

Map Group Load Modules

Code for generated map groups is stored in these modules. One module is created for each device type supported by the map group. The module names begin with the map group name and are suffixed by a two-character device type.

Processing Load Module

Processing statements, map edits, and literals are stored in the Processing Load Module. Processes, statement groups, and special EZE functions are stored in the Application Processing Table (APT). Processing statements from both processes and statement groups are in the Process Statement Table (PST). The literal pool (LIT) holds all literals used in processing statements. The Item Edit Table (IET) holds specifications regarding built-in map edits used by the application. An Execution SQL Process Table (ESPT) contains SQL process information, and the Execution SQL Record Table (ESRT) holds data concerning SQL records. The processing load module's name is padded with dollar signs ($) and suffixed by the letter "P" (TZ01A becomes TZ01A$$$P).

SQL Statement Execution Load Module

This module includes statement IDs from SQL statements used in SQL/DS extended dynamic mode and DB2 static mode in the SQL Statement ID table (SSIT). It also contains abbreviated information necessary to generate dynamic SQL calls used by the application in a shortened Preparation SQL Record Table (PSRT). The module name is padded with dollar signs ($) and suffixed with an "B" (TZ01A becomes TZ01A$$$B).

SQL Statement Preparation Load Module

Both the Preparation SQL Process Table (PSPT) and the full Preparation SQL Record table (PSRT) are stored in this load module. The module name is padded with dollar signs ($) and suffixed with an "A" (TZ01A becomes TZ01A$$A).

Table Load Modules

The definitions for data tables are stored in these modules. The name of the load module is padded with zeros (0) and suffixed with a "T" (TZ01T1 becomes TZ01T10T). Make sure your naming convention doesn't get you into trouble here!

Even though CSP calls them load modules, the output from a generation is not executable. The CSP/AE environment must be in place to execute CICS applications.

CSP APPLICATION EXECUTION (CSP/AE)

CSP/AE (Cross System Product—Application Execution) is the complementary product to CSP/AD (Cross System Product—Application Development) that allows the execution of generated CSP applications. Some systems (most notably MS-DOS and IMS) only allow execution of CSP/AE modules that have been generated in another environment. Most environments use the XSPE transaction to execute CSP/AE. Some of the syntax to execute CSP/AE under CICS is in Figure 14.6. For full syntax description, see the CSP manuals for your system.

The ALF (Application Load File) name (alfnam) identifies the ALF containing the application's generated output. If you leave it out, CSP defaults to an ALF name of FZERSAM. The application name (applnam) is the application to be executed. SEG should be specified in CICS environments to cause an application

```
XSPE [alfnam.]applnam
    SEG
    DMODE=D
    RT=trid
    TSMS
    NOTXA
```

Figure 14.6.   Executing CSP/AE.

generated for either segmented or unsegmented to run segmented.

DMODE=D specifies that dynamic SQL execution is used. If omitted, static SQL execution is assumed. RT=trID names the transaction ID to be executed upon exit from CSP/AE. This is useful if you are attempting to keep the user inside of a set of menus. TSMS causes data stored between segments to be placed in CICS Main Temporary Storage rather than Auxiliary Temporary Storage. Using TSMS is a good idea if your installation has a surplus of memory; otherwise, leave it out. NOTXA indicates that the module will not support 31-bit addressing. NOTXA should be used any time an application will be calling modules that cannot support 31-bit addresses (OS/VS COBOL for instance).

CSP/AD UTILITIES

CSP provides several utility functions and the ability to view and modify records under the Utilities and File Maintenance screens. To get to the utilities menu, choose facility 5 (Utilities and File Maintenance) from the CSP/AD FACILITY SELECTION (main menu) panel. See Figure 14.7. The items listed in the left column are different CSP/AD Utilities that you may execute. The two items in the right column allow entry into the File Maintenance function in read-only mode (View) or with the intent to alter records (Change).

1. Member Print to print individual MSL members
2. Directory Print to print all members from an MSL
3. Export an MSL member (or group of members) to an ALF or to an External Source Format file
4. Import an MSL member (or group of members) from an ALF or from an External Source Format file
5. Copy MSL member(s)
6. Rename MSL member(s)
7. Delete MSL member(s)
8. Change data files using CSP record definitions
9. View data files using CSP record definitions

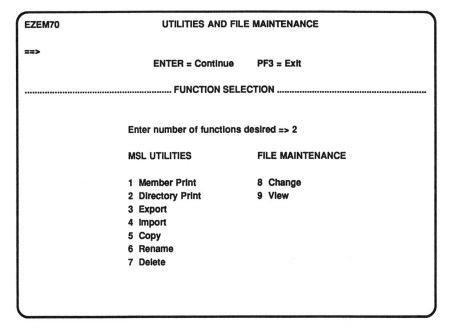

Figure 14.7. Utilities and file maintenance.

Member Print Utility

This facility allows the printing of MSL members. This is the same function invoked by using the option "P" from the LIST PROCESSOR panel, but it allows more options (Figure 14.8). List the name of the member to be printed and choose any other options you want (you may not print individual data items; they may be printed along with the objects that use them). Choosing the Cross-Reference causes the Cross-Reference display to appear (Figure 14.9). Fold print forces all lower-case letters to be converted to upper case during printing. Lines per page may be any value between 20 and 999 (inclusive). This allows a small amount of customization to fit available printers. You may print up to 99 copies (ignored when output is to a file). If MSLs are concatenated, only the member found in the highest level MSL is printed. Be sure to tell CSP if the printer involved is DBCS (Double Byte Character Set) capable.

```
EZEM71                          UTILITIES

==>
                    ENTER = Continue      PF3 = Exit

............................................ MEMBER PRINT ...........................................


    Enter name of member to be printed => xxxx

    Enter output options:

            Cross Reference       => NO
            Fold Print            => YES
            Lines per page        => 55
            Copies                => 1
            DBCS Printer          => YES
```

Figure 14.8.   Member print utility.

```
EZEM75                          UTILITIES

==>
                    ENTER = Continue      PF3 = Exit
                        Name = xxxxx
.......................................... CROSS REFERENCE ......................................


    Print data item definitions?       => NO
    Include member types: Data Items => YES   Applications      => YES
                          Processes  => YES   Statement Groups => YES
                          Records    => YES   Tables            => YES
                          Maps       => YES   Map Groups        => YES
                          PSBs       => YES
```

Figure 14.9.   Cross-reference display.

Cross-Reference Display

If Cross-Reference is selected from either the Member Print or Directory Print utility, the following screen displays. For Directory Print, this screen appears once for each application in the MSL. The cross-reference listings can be very helpful when debugging or maintaining an application.

To print data item definitions along with the members that reference the data item, mark the "Print data items definitions" with "yes." The "Include Member Types" choices govern which types of members will be printed. CSP's default is to print all available types of members. Mark any that are not desired "no" in the space provided.

Directory Print Utility

This utility lists contents of all members from a selected MSL. It does not provide a list of members as its name implies (Figure 14.10). Like the member list utility, specify whether you want to

```
EZEM72                          UTILITIES

==>
                        ENTER = Continue      PF3 = Exit

.................................. DIRECTORY PRINT ..................................


        Enter MSL number for print      => XXXX

        Enter output options:

                Cross Reference        => NO
                Lines per page         => 55
                Copies                 => 1
```

Figure 14.10.   Directory print.

view the cross-reference choices, the number of lines per page desired, and how many copies should be printed.

Export Utility

Export is used to move things from an MSL to an ALF or to a file in External Source Format (see Chapter 15). The export function is an essential part of "porting" an application between environments. You must make sure that a name being EXPORTed does not match anything in the target ALF file (to do, use the ALF utility described in Chapter 16). If concatenated MSLs are involved, the first MSL containing a member with the specified name is used.

The EXPORT illustrated by Figure 14.11 and the IMPORT shown by Figure 14.12 may be used to copy the MS10A (Message File Utility) application into your read-write MSL. First, make sure that the MSL named "UTILMSL" is listed at the top of the

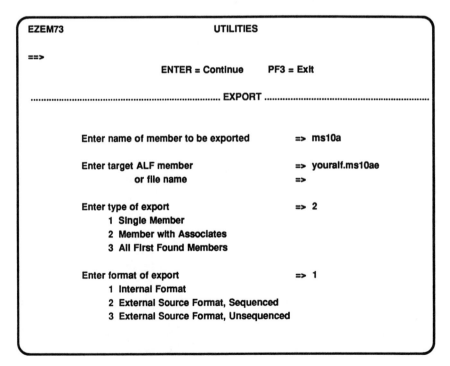

Figure 14.11. Export utility.

```
EZEM74                              UTILITIES

==>
              ENTER = Continue    PF3 = Exit      PF5 = View Messages

.......................................... IMPORT ........................................


           Enter source ALF member        => youralf.ms10ae
                  or file name            =>

           Replace duplicate members       => YES

           Selective Import requested      => NO

           Selective import list option    => 1
               1  All
               2  Duplicates Only
               3  Non-Duplicates Only

           On error, import valid members  => YES
```

Figure 14.12. Import utility.

MSL SELECTION panel (use the "=MSl" command to get there). Return to the EXPORT panel by using the "=EXport" command. The panel in Figure 14.11 shows how you might EXPORT the application named MS10A (first one found in your MSL application wins) to the ALF named (youralf), using the member name listed (ms10ae). If you omit the ALF name, FZERSAM is assumed. Please notice that the ALF member name chosen was the application name suffixed by an "E" (for export). The use of a suffix "E" is an unofficial standard for exported files. Choose a file name (ddname) when exporting External Source Format (see format below). The file name must be allocated so that your CSP/AD session can find it.

Type of export tells CSP the scope of the export. Single Member means just what its name implies: only the named member is exported. Member With Associates is the most common way to export. Exporting the member with associates includes all processes, tables, and map groups involved with an application. All First Found Members should be used very carefully. It causes all members from all concatenated MSLs to be exported on a first-found basis. If a member is in more than one MSL, only the first MSL's member is exported (first found).

Format of export is very important. If you are exporting for the purpose of moving something from one MSL to another, choose Internal Format (as in the example). If you wish to export using the External Source Format (see Chapter 15), you may choose to generate records with sequence numbers in columns 73–80 (Sequenced) or without sequence numbers (Unsequenced). Since you are unlikely to be creating a stack of cards that might then be dropped, boldly forgo the sequence numbers.

Import Utility

Import is used to bring an EXPORTed member into a R/W MSL (Figure 14.12). If the member being imported was EXPORTed to an ALF previously in internal format, specify the ALF name, followed by a period (.), followed by the member name on the ALF (ms10ae in Figure 14.12). If the member being imported is in External Source Format (see Chapter 15), specify the file (ddname) containing the data. This file (ddname) must already be defined to your CSP/AD session. Specify REPLACE to copy over an existing member with the results of the import (make sure you have not made a mistake).

Selective import allows you to view a list of objects (not illustrated here) contained in an exported member (when using internal format). You may choose from all of the objects (duplicates will be marked), only duplicates, or only non-duplicates. You may rename duplicate objects either in the existing MSL or coming from the export as your needs dictate. The final choice (On Error, Import Valid Members) is used when importing members stored in External Source Format. The Import utility validates the CSP syntax as it imports External Source Format code. If "On Error, Import Valid Members" is marked yes, Import will write individual members to the MSL as they are verified. If the field is marked no, Import must hold all members in memory until the last one passes (this will run much slower than specifying yes).

Copy Utility

The copy utility (Figure 14.13) is used to copy a member from one MSL into your read-write MSL. The MSLs must both be allocated and be listed in the current list of MSLs used by your CSP/AD session. A list of member names may be obtained using the LIST PROCESSOR. Copy is also an option under the LIST PROCESSOR.

Enter the name of the source member to be copied and the number of the source MSL (from the concatenation list displayed on the MSL SELECTION panel). If no MSL number is entered, the first member found in the current MSL concatenation is copied.

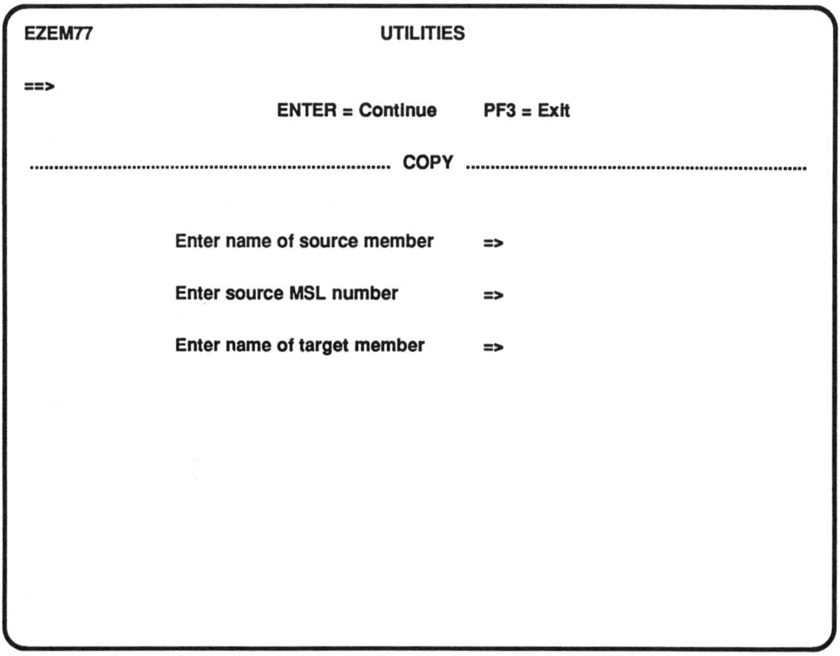

```
EZEM77                          UTILITIES

==>
                    ENTER = Continue      PF3 = Exit

............................................. COPY .............................................


          Enter name of source member      =>

          Enter source MSL number          =>

          Enter name of target member      =>
```

Figure 14.13. Copy utility.

The name of the member in the target MSL may be different from the source's; simply key in the new name. For maps, both the map group and map name must be entered. If you would like to use the source's name for the target, enter an equal sign (=). For maps, the map group name, map name, or both may be equal signs (= =).

Be careful! Copying applications *does not* copy the processes and statement groups that are associated with the application. Copying a record copies local data items, but it does not copy the GLOBAL data items that are used by the record (only the name in the record and the fact that it is GLOBAL).

Rename Utility

The rename utility (Figure 14.14) allows renaming of individual members in the read-write MSL. A list of members may be obtained using the LIST PROCESSOR. Rename is one of the options under the LIST PROCESSOR.

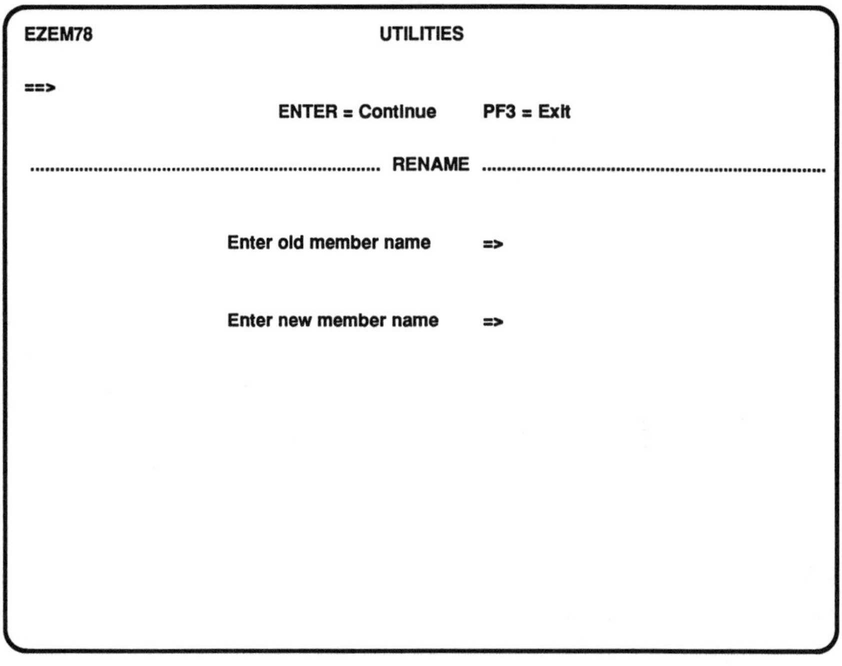

```
EZEM78                        UTILITIES


==>
                     ENTER = Continue     PF3 = Exit

................................................ RENAME ................................................


              Enter old member name     =>


              Enter new member name     =>
```

Figure 14.14. Rename utility.

Enter the existing member name and the new name, then press ENTER. *Caution!* Changing a member name DOES NOT alter the name anywhere it appears in applications, processes, maps, or record definitions. This change may invalidate one or more other objects. Check CSP documentation and Chapter 7 for information on the LIST PROCESSOR's Change option.

Delete Utility

This utility (Figure 14.15) allows deletion of individual members from the read-write MSL. A list of members may be obtained using the LIST PROCESSOR. Delete is one of the options under the LIST PROCESSOR.

Enter the existing member name, then press ENTER to delete the member. CSP will *not* ask if you mean to do this; the member is deleted instantly. (Have a nice day!) Be careful that members being deleted are not being used in any current applications, processes, maps, or record definitions.

```
EZEM79                        UTILITIES

==>
                   ENTER = Continue      PF3 = Exit

................................................. DELETE .................................................


               Enter name of member
                 to be deleted        =>
```

Figure 14.15.   Delete utility.

CSP FILE MAINTENANCE

CSP/AD's built-in file maintenance allows a user to view and/or modify data from indexed, relative, or serial files. The file maintenance is based upon the record definition stored in an MSL. While your installation may already have products used to view and edit file data, this product has two principal benefits. First, this is an excellent tool for testing CSP record definitions. Second, you may view/alter files without leaving CSP/AD.

Using the File Maintenance facility is a two-screen process. First the CSP record and file name are specified. Then, another panel lets you choose which fields to view/change and from which records. Of course, you may only view and alter files that you normally have security access to. See Figure 14.16.

Enter the name of the record definition to be used. This name must be found in the current MSL concatenation list. If the record definition's file name has been allocated, CSP displays that value as a default. A question mark (?) may be entered to

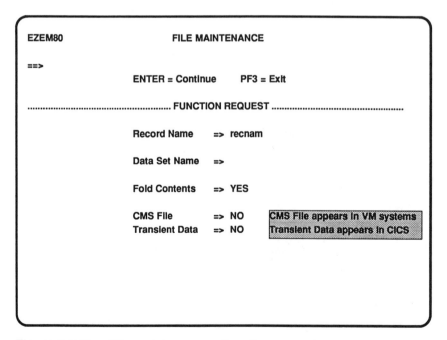

Figure 14.16. File maintenance—Function request.

verify the default file name. If the record definition may be used for multiple files, simply key in the name of another file to be viewed or altered. Please note that the value in Data Set Name is dictated by the environment. CICS requires a ddname, TSO requires a DSN, and VM requires a real DSN for VSAM files or a FILE ID for other file types. Consult your system documentation for specifics.

Tell CSP to convert lower-case letters to upper case if that is necessary for the terminal you are using.

The bottom two fields apply only to specific environments and may not show up on your display. CMS file is used under VM to tell CSP that a serial file is not a VSAM ESDS. Transient Data is used under CICS for serial files only (names are limited to 1–4 characters). After keying in the record information, press ENTER to view the FILE MAINTENANCE ITEM CONTENTS DISPLAY (Figure 14.17). Whether you specified option 8 (Change) or option 9 (View) from the UTILITIES AND FILE MAINTENANCE FUNCTION SELECTION panel, the panels used for file

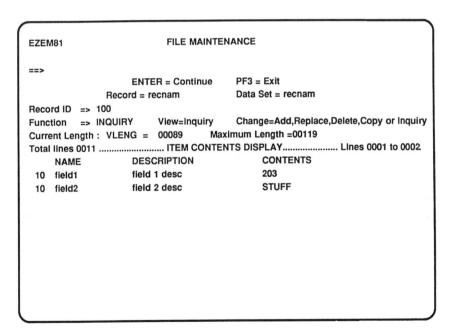

Figure 14.17. File maintenance—Item contents display.

maintenance are the same. The second panel displayed allows the user to specify which fields are to be viewed/changed and to select the individual records.

The first time this panel is displayed, the message "Enter specific item names or press enter for select list" is displayed. You may either enter the names of data items you wish to view/ change (if you know them), or you may press ENTER to see a list of all items in the record description (usually the best course). If you wish to view the entire record, leave things as they are displayed and continue to select records. However, CSP lists all data items in the definition, including group items. This may clutter up your display with data both from the group items and the detail items. Mark which fields you truly wish to see by placing an "S" in front of the field name. Don't press ENTER until all fields have been marked! To alter this list you must exit to the FILE MAINTENANCE FUNCTION REQUEST panel and start over. (I know—you would have been more user friendly!)

To choose a record (for indexed and relative files), enter a key value or RRN in the Record ID field. For serial files, or for sequential processing of indexed or relative files, CSP advances one record at a time each time you press ENTER.

Next, choose the function desired. The function names (except COPY) are similar in name and function to the CSP Process Options allowed for these records. Inquiry is used to read records. Add is used to create new records. Replace is used to modify existing records. Delete is used to delete records (not be supported by all file types). Copy is used to create a new record like an existing one. Indexed and Relative records are read matching the key/RRN specified, then the record is presented for modification. Copy for sequential records simply sets up the last record processed as the record to be modified. Under View mode you may only perform inquiries. Even under Change mode it makes sense to look at something before you change it.

Variable-length records display slightly different forms of the FILE MAINTENANCE ITEM CONTENTS DISPLAY panel based upon the definition (depends on use of length field or occurs field). Current Length (when displayed) may be modified for variable-length records when ADDing or REPLACEing or COPYing. Number of Occurs (when displayed) may be modified for variable-length records when ADDing or REPLACEing or COPYing.

The four columns displayed at the bottom of the panel include level number of data item, data item name, description of data item, and contents. Contents are shown 20 characters at a time, over as many lines as are necessary. If all data items are not in the current display, scroll to find the data items(s) desired. This is one of those panels where remembering that PF9 may be used to hide the top portion of the screen is useful. When ADDing, REPLACEing, or COPYing, the contents of data items may be modified.

Be careful! The file maintenance utility assumes that you know exactly what you are doing and that any record deletions or changes are intentional. No warnings take place; the action specified is simply done without question. (Once again, have a nice day!)

BATCH EXECUTION OF CSP/AD

The CSP/AD product may be executed in batch mode in most environments. The *CSP/AD Developing Applications* manual documents the batch commands, while the individual *CSP/AD Administration* manual documents the process for each host environment. Most of the commands are identical in function to activities available on line, but some commands are unique to the batch environment. Some illustrations in this chapter are for the MVS, but all commands work the same regardless of environment. To execute the CSP/AD commands in batch, the correct MSL and ALF file names must be used, and the syntax (Figure 14.18) must be followed exactly.

Command syntax is unforgiving (as is most syntax) but basically pretty simple. Please note that commands are continued from line to line without continuation marks and that the semicolon (;) must be used to delimit (end) each command. Illustrations of various batch commands follow. All of the commands work the same way in batch as they do interactively.

The command in Figure 14.19 is unique to the batch environment and should be followed by a batch GENERATE command. If used, ASSOCIATE tells CSP which CSP record definition file

- **Comments begin with /* and end with */**

- **Multiple commands may share a line**

- **Commands may be continued from line to line**

- **Parameters are keyword**

- **Case is unimportant**

- **Use a semicolon to delimit commands**

- **Sample JCL streams follow the commands**

Figure 14.18.   Batch CSP/AD syntax.

```
ASSOCIATE FILE(filenam)
     SYSNAME('real.system.name')
     FILETYPE(xxxx)
         PCBNO(nn);
```

Figure 14.19.   ASSOCIATE syntax.

name matches up to which real file name on the system. If ASSO-CIATE is not used before a GENERATE, CSP defaults the association last SAVEd as part of a generation or the system defaults. Each GENERATE used the previous ASSOCIATE command's output. If two (or more) generations are done, it is often desirable to include ASSOCIATE commands before each one. It is common to omit the ASSOCIATE command and generate with the defaults.

FILE names the file name used in the application being generated. SYSNAME is for the file name on the host system (format is environment dependent). FILETYPE is a good idea when a file type may be ambiguous (serial files in many environments). FILETYPE is also dependent upon the target environment. The following are the choices available:

File Type	Description
CMS	VM/CMS file
VSAMCMS	VM VSAM ESDS
DATABASE	DPPX DTMS database
FILE	DPPX sequential file
VSAM	VSAM ESDS used in MVS batch or IMS BMP
GSAM	GSAM database for IMS BMP or MVS batch
SEQ	MVS sequential file used in MVS batch or IMS BMP (default for MVS batch and IMS BMP)
MMSGQ	Multi-segment message queue used in IMS/TM or IMS BMP
SMSGQ	Single-segment message queue used in IMS/TM or IMS BMP
TRANSIENT	CICS Transient Data Queues
VSAMCICS	CICS ESDS

IMS applications may also need the PCBNO option. This is the number of the PCB in the PSB member for either the message queue or GSAM database associated with the file. If omitted, PCBNO defaults based upon the PSB definition and use of the file. SMSGQ and MMSGQ files default to "0" for input, "1" for output. GSAM files default to the first PCB.

Copy works just like online (Figure 14.20). Enter the source member name (name1), the target member name (name2 or "="), and the name/number of the MSL where the source may be found. Only one member may be copied per statement, however, multiple copies are allowed.

Figure 14.21 shows the delete member. As with the online

```
COPY MEMBER(name1)
     NEWNAME(name2)
     MSL(n or name);
```

Figure 14.20. COPY syntax.

```
DELETE MEMBER(name);
```

Figure 14.21. DELETE syntax.

```
DIRECTORY MSL(n or name)
          CROSSREF(Y or N)
          COPIES(ncopies)
          LINES(nlines)
          REFTYPE(xxxx,yyyy)
          PRINTITEM(Y or N);
```

Figure 14.22. DIRECTORY list syntax.

utility, enter the name of the member to be deleted from the read/write MSL. For maps, specify both map and map group names.

The Directory Print capability (Figure 14.22) is probably more reasonable to execute in batch mode than online. Since it will list all members from the specified MSL, it may take considerable time (and kill many trees if printed!). Specify the MSL number or name, whether a cross reference is desired, the number of copies, lines per page, types of members to be listed, and whether individual data items definitions are to be listed. MSL points to either the name or number (1–6) of the MSL to be printed from. CROSSREF creates a cross-reference of any members that reference other members. COPIES may be from 1 to 99, the default is 1. LINES may be from 20 (largest heading supplied by CSP) and 999, the default is 55. REFTYPE may be one or a list of member types (separated by commas inside parentheses) using similar type codes to those shown by the LIST PROCESSOR:

Type	Description
APPL	Application
ITEM	Data item
MAP	Maps
MAPG	Map groups
PROC	Processes
PSB	PSBs
RECD	Records
SGRP	Statement groups
TBLE	Tables

PRINTITEM indicates whether the data item definitions are to be printed along with the members that reference them. The default is no.

EXPORT (Figure 14.23) is probably one of the most frequent reasons to use batch commands. This is used when copying many items from one MSL to another. It is also useful when EXPORTing External Source Format for editing or transfer to the PAS (Programmable Workstation). Member lists the name of the MSL member to be exported. Specify either SERIAL to name a sequential file's ddname/dlbl for output, or FILE to name an ALF's file name (ddname or DLBL). When specifying an ALF, you may also specify TARGET to choose the member name on the output ALF. It is most common to use ALFs for EXPORTing in INTERNAL format and sequential files for EXPORTing in EXTERNAL format. Under MVS, the sequential file being EXPORTed and later IMPORTed may be the member of a PDS

```
EXPORT MEMBER(name)
     SERIAL(ddname) or FILE(alfdd)  TARGET(alfmbr)
     TYPE(MEMBER or ASSOCIATES or ALL)
     FORMAT(INTERNAL or EXTERNAL or EXTERNALU);
```

Figure 14.23. EXPORT syntax.

(Partitioned Data Set) as long as the member is identified in the ddname allocation. Be careful to name output targets so that the output from a generation is not overwritten (common practice is to append member name with an "E" for export). If TARGET is omitted, the source member name is used. If the source member's name is longer than eight (8) characters or contains underscores (_) or hyphens (–), either TARGET or SERIAL must be specified.

TYPE describes the scope of the EXPORT. MEMBER means that only the named member is to be exported. ASSOCIATES means that all members associated with the source member will be exported on a first-found basis (first occurrence in the MSL concatenation is exported). ALL should be used with caution; it means to export all members from all MSLs in the current MSL concatenation on a first-found basis. Format describes the future use of the EXPORTed member(s). INTERNAL means that records suitable for later IMPORT are to be generated; it is intended for moving MSL members from one library to another. EXTERNAL and EXTERNALU show that External Source Format is to be used. This format is useful when the exported member(s) will be updated using an editor or shipped to a PWS (Programmable Workstation) for processing. EXTERNAL creates 80-byte records with sequence numbers in columns 73–80 (in case you drop that deck). EXTERNALU creates 80-byte records with blanks in columns 73–80. It is usually simpler and cleaner to use EXTERNALU.

IMPORT (Figure 14.24) is the reverse of EXPORT. It allows you to copy members back into MSLs using either INTERNAL format files or EXTERNAL SOURCE FORMAT. To IMPORT from a sequential file (or PDS member in MVS), specify SERIAL and name the ddname/dlbl where the EXPORTed data may be found. To import from an ALF, specify FILE and name the

```
IMPORT      SERIAL(ddname)
            or FILE(alfname) SOURCE(alfmbr)
         REPLACE(Y or N)
         BYPASS(Y or N);
```

Figure 14.24. IMPORT syntax.

ddname/dlbl for the ALF; also specify SOURCE to identify the member in the ALF to be IMPORTed. CSP will detect whether the data being IMPORTed is in INTERNAL or EXTERNAL format and process it accordingly. It is most common to use sequential files when EXPORTing and IMPORTing EXTERNAL Source Format data and to use ALFs when EXPORTing and IMPORTing INTERNAL format data.

If you want IMPORT to stop when an IMPORTed member has the same name as an existing member in the read-write MSL, specify REPLACE(N). Specify REPLACE(Y) to copy over any like-named members in the MSL. REPLACE(N) is the default.

When importing from External Source Format, CSP validates the incoming statements and definitions. Specify BYPASS(Y) to skip invalid definitions and continue the IMPORT (the default). To cause the IMPORT to halt when an error is found, specify BYPASS(N).

Batch execution of GENERATE (Figure 14.25) is a good idea for many reasons, primarily relieving the online system of the

```
GENERATE MEMBER(name)
    FILE(alfdd)
    SYSTEM(CICS/OS or CICS/DOS or MVSBATCH
            or VSEBATCH or TSO or CMS or OS/400
            or DPPX or ANY)

    RESET(Y or N)
    PRINT(Y or N)
    FOLD(Y or N)
    REF(Y or N)
    TABLES(ALL or NONE)
    MAPS(ALL or NONE)
    SQLVALID(Y or N)
    LOCVALID(Y or N)
    EXECMODE(1 or 2 or 3 or 4)
    TRANSID("txid")
    COBOLSYS(IMS/VS or IMSBMP or MVSBATCH)
    SAVE(Y or N);
```

Figure 14.25. GENERATE syntax.

processing required to GENERATE. Specify the MEMBER name to be generated, the name of the ALF to be generated into (FILE, defaults to FZERSAM), and the target SYSTEM. Other options may be specified as desired or needed. Defaults have been underlined in Figure 14.25. See GENERATION earlier in this chapter for more detailed information on the generation process and its options. RESET indicates that previously saved generation options are to be deleted and ignored. Options are defaulted unless overridden by the GENERATE, ASSOCIATE, or SETGEN commands. SAVE causes CSP to save the current generation options in the MSL for future use. PRINT causes a listing of generation source code (useful for debugging). FOLD causes the printed listing's data to be forced to upper case. REF indicates that a reference file is to be created (also useful for debugging). TABLES controls whether the tables associated with this application are also generated. MAPS controls whether the map groups associated with this application are also generated. SQLVALID requests that CSP perform anticipatory SQL syntax checking that duplicates what will be done later by the SQL preprocessor. LOCVALID ensures that local data items match global data items of the same name. EXECMODE is used in CICS and DPPX systems to control segmentation: 1 = Nonsegmented, 2 = Segmented, 3 = Either, 4 = Single Segment (defaults to 2 for IMS, 1 for all other systems). TRANSID specifies the segmented transaction ID used for CICS and DPPX segmented transactions.

COBOLSYS is used when creating COBOL source code as part of the generation. Currently, you may run COBOL CSP programs in the MVS batch, IMS/TM, and IMS BMP environments. GENERATE will create the COBOL source code for the three COBOL execution environments as specified (MVSBATCH, IMS/VS, IMSBMP). COBOL generation may be performed only in a batch job. The COBOL programs must be compiled and linked in the normal fashion, and at execution time the CSP/370 Runtime Services modules must be installed for execution to occur. This is the manner in which future CSP execution under MVS will take place once CSP/AD Version 4 and the next release of CSP/370 Runtime Services are available.

The MSL command (Figure 14.26) is used to change the concatenation sequence for MSLs during the batch execution. This is useful when performing lots of EXPORTs and IMPORTs. The

```
MSL        M(rwmsl)
           ROMSL(romsl1,romsl2);
```

Figure 14.26.   MSL syntax.

"M" names the MSL to be the read-write MSL for subsequent commands. The "ROMSL" list specifies the read-only MSLs and their respective place in the MSL concatenation. Remember, read-write MSLs are not required by all operations and sometime it is useful to have only read-only MSLs. The MSL names in this command are ddnames/dlbls and must have been allocated as part of the batch execution.

The PRINT utility (Figure 14.27) works the same way in batch as it does online. Specify the member name (except data item names) to be printed, whether you want the output converted to upper case, the number of lines per page, and the number of copies. Printing an application implicitly prints all associated objects. Specifying CROSSREF(Y) tells CSP to cross-reference all member names found in the application. If CROSSREF(Y) is used, specific types of members may be singled out for cross-referencing using the REFTYPE list. Enter the type(s) of members to be cross-referenced (separated by commas, surrounded by parentheses).

Type	Description
APPL	Application
ITEM	Data item
MAP	Maps
MAPG	Map groups
PROC	Processes
PSB	PSBs
RECD	Records
SGRP	Statement groups
TBLE	Tables

```
PRINT MEMBER(name)
    FOLD(Y or N)
    CROSSREF(Y or N)
    LINES(nlines)
    COPIES(ncopies)
    REFTYPE(xxxx,yyyy);
```

Figure 14.27.   PRINT MEMBER syntax.

```
RENAME MEMBER(name1)
    NEWNAME(name2);
```

Figure 14.28.   RENAME syntax.

Rename (Figure 14.28) is identical to renaming members online. Specify the old member name and the new member name. For maps, both map group and map names are needed. The new name must not already exist in the read-write MSL.

The SET command (Figure 14.29) allows you to set a return code that will cause the CSP/AD batch processing to stop. Enter a number between 0 and 16 or a STOPON keyword depending upon the severity you wish to allow. If a return code greater than that in the SET command occurs, processing will stop.

Return Code	Usual Meaning
0	Everything is wonderful
4	CSP issued a warning
8	Error that may be survivable
12	Severe error
WARN	Same as 4
ERR	Same as 8
SEV	Same as 12

SET STOPON(rcode);

Figure 14.29. SET STOPON syntax.

SETGEN (Figure 14.30) is used only in batch. It allows specification of options for generation of applications, tables, and maps. Options specified are saved in the MSL for the next generation of the same member only if you also specify SAVE(Y) on the GENERATE command. SETGEN is used only for the next GENERATE command; two GENERATEs require two SETGENs. If SETGEN is omitted, GENERATE uses saved options from the MSL or the system defaults (depending upon the setting of the RESET option in the GENERATE command). This is useful if an online generation has not been saved specifying these values.

GENERATE specifies whether modules in other keywords for this SETGEN command (GROUP, TABLE, CLIST) will be turned "on" or "off" as part of the defaults. CLIST, DDS, GROUP, and TABLE reflect the options to generate CLIST for DPPX systems, DDS modules for OS/400, map GROUPs, and data TABLEs. If TABLE is specified, whether the table is to be SHARED and KEEP after use may be turned on as well.

For SQL applications, the generation of ANSI Static modules (ANSIMOD), DB2 Static Modules (DB2MOD), and SQL/DS AC-

---

```
SETGEN  GENERATE(Y OR N)
        CLIST
        DDS
        GROUP(grpnam)
        TABLE(tblnam)
            SHARED(Y or N)      KEEP(Y or N)
        SQLMOD(accmod)
            USER('userid')  PASSWD('pswd')
            BLOCK(Y or N)
        DB2MOD(dstatmod)
        ANSIMOD(astatmod);
```

Figure 14.30. SETGEN syntax.

CESS (SQLMOD) modules may be turned on or off, and specific names may be given to the respective modules. If turning on SQL/DS ACCESS modules (SQLMOD), the userID, password, and use of blocking is also an option.

NON-DB2 JCL FOR BATCH CSP/AD

Figure 14.31 shows JCL that may be used to execute the CSP/AD product in batch. If you are running in a non-MVS environment, see the CSP Administration manuals for your environment. This JCL is for use in non-DB2 situations. If DB2 is used, see the JCL in Figure 14.32.

The program executed by this JCL is actually CSP/AD (DCGBINIT). You will probably need to find out what the first one or two nodes of the CSP files used in your installation are from the

```
//xxx JOB
//stepnam    EXEC  PGM=DCGBINIT
//STEPLIB    DD  DSN=CSP322.AELOAD,DISP=SHR
//               DD  DSN=CSP322.ADLOAD,DISP=SHR
//DCAEZED    DD  DSN=CSP322.EZEMSG,DISP=SHR
//DCAMAPD    DD  DSN=CSP322.FZEMAPDS,DISP=SHR
//DCAHECD    DD  DSN=CSP322.FZEMSG,DISP=SHR
//DCATESD    DD  DSN=CSP322.FZETUTOR,DISP=SHR
//usrrwmsl   DD  DSN=user.rw.msl,DISP=OLD
//romsl1     DD  DSN=user.ro.msl1,DISP=SHR
//romsl2     DD  DSN=user.ro.msl2,DISP=SHR
//DCAWORK    DD  DSN=CSP.USER.DCAWORK,DISP=OLD
//EZECOUT    DD  SYSOUT=*,DCB=(RECFM=FBA,LRECL=121,BLKSIZE=1210)
//EZECIN     DD  *
  /**** commands go here ****/
    PRINT MEMBER(myapp);
/*
//FZERSAM    DD  DSN=user.alf,DISP=OLD
//EZEPRINT   DD  SYSOUT=*,DCB=(RECFM=VBA,LRECL=137,BLKSIZE=141)
//SYSPRINT   DD  SYSOUT=*
//DCAPARM    DD  *
M=usrrwmsl ROMSL=(romsl1,romsl2)    CMDIN
/*
//
```

Figure 14.31. CSP/AD MVS JCL (non-DB2).

```
//xxx JOB
//stepnam   EXEC PGM=IKJEFT01
//STEPLIB   DD DSN=CSP322.AELOAD,DISP=SHR
//          DD DSN=CSP322.ADLOAD,DISP=SHR
//DCAEZED   DD DSN=CSP322.EZEMSG,DISP=SHR
//DCAMAPD   DD DSN=CSP322.FZEMAPDS,DISP=SHR
//DCAHECD   DD DSN=CSP322.FZEMSG,DISP=SHR
//DCATESD   DD DSN=CSP322.FZETUTOR,DISP=SHR
//SYSTSPRT  DD SYSOUT=*
//SYSTSIN   DD *
  DSN   SYSTEM(DSN)
  RUN PROGRAM(DCGBINIT)      +
   PLAN(DCGPLAN)             +
   LIB('CSP322.AELOAD')
/*
//usrrwmsl  DD DSN=user.rw.msl,DISP=OLD
//romsl1    DD DSN=user.ro.msl,DISP=SHR
//romsl2    DD DSN=user.ro.msl,DISP=SHR
//DCAWORK   DD DSN=CSP.USER.DCAWORK,DISP=OLD
//EZECOUT   DD SYSOUT=*,DCB=(RECFM=FBA,LRECL=121,BLKSIZE=1210)
//EZECIN    DD *
  /**** commands go here ****/
   PRINT MEMBER(myapp);
/*
//FZERSAM   DD DSN=user.alf,DISP=OLD
//EZEPRINT  DD SYSOUT=*,DCB=(RECFM=VBA,LRECL=137,BLKSIZE=141)
//SYSPRINT  DD SYSOUT=*
//DCAPARM   DD *
M=usrrwmsl ROMSL=(romsl1,romsl2)    CMDIN
/*
//
```

Figure 14.32.   CSP/AD MVS JCL (DB2 okay).

CSP administrator (in fact, they may have working JCL for you already). The three DD statements, usrrwmsl, romsl1, and romsl2, are optional, and you may use any ddnames you want. These three ddnames are referred to by later control statements. Printed status concerning the success/failure of your commands will be sent to the ddname EZECOUT. EZECIN is where the commands listed on the previous pages go. Don't forget to delimit them with semicolons (;). FZERSAM is the name of the default ALF. If other ALFs will be used, you must include DD statements for them too. EZEPRINT is where output from the PRINT commands will be sent. DCARPARM is how you tell CSP/AD about the original

MSL concatenation. In Figure 14.31, "M=usrrwmsl" tells CSP that the ddname "usrrwmsl" will serve as the read-write MSL for processing. "ROMSL=(romsl1,romsl2)" tells CSP that the ddnames of two read-only MSLs will be used with "romsl1" being in front of "romsl2" in the concatenation order. The concatenation order may be altered dynamically by the MSL command as long as the ddnames are there to support it. You may use any ddnames that make you feel good. "CMDIN" tells CSP that the input commands will come from the EZECIN ddname. You may alter this by entering "CMDIN=myddnam" to redirect the input commands. Many other (less frequently used) options are possible for input to DCAPARM. See the Administration manual for your environment for details.

DB2 JCL FOR BATCH CSP/AD

Figure 14.32 shows JCL that may be used to execute the CSP/AD product in batch when DB2 is involved. If you are running in a non-MVS environment, see the CSP Administration manuals for your environment. This JCL is for use in DB2 situations, for instance, when generating applications that require static load modules.

The program executed by this JCL is the TSO Terminal Monitor Program (IKJEFT01). You will probably need to find out what the first one or two nodes of the CSP files used in your installation are from the CSP administrator (in fact, they may have working JCL for you already). This JCL is identical to that shown in Figure 14.31 except for additional statements required because DB2 is being used. The SYSTSPRT ddname is where messages concerning success/failure of your attempt to run CSP/AD in batch with DB2 will be found. SYSTSIN is the set of commands that tell the TSO program what to do. "DSN SYSTEM(DSN1)" tells TSO which DB2 subsystem to connect to. You will probably have to replace "DSN1" with a correct DB2 subsystem name for your shop (get it from your DBA). "RUN PROGRAM(DCGBINIT) PLAN(DCGPLAN) LIB('CSP322.AELOAD')" tells TSO which program to run (DCGBINIT again), what application plan it uses (DCGPLAN), and where the load module for DCGBINIT may be found (in this case "CSP322.AELOAD"). Any or all of the three names may have to change for your environment. The plus signs (+) in the example are continuation marks required because the command was entered over three lines.

The three DD statements, usrrwmsl, romsl1, and romsl2, are optional, and you may use any ddnames you want. These three ddnames are referred to by later control statements. Printed status concerning the success/failure of your commands will be sent to the ddname EZECOUT. EZECIN is where the commands listed on the previous pages go. Don't forget to delimit them with semicolons (;). FZERSAM is the name of the default ALF. If other ALFs will be used, you must include DD statements for them too. EZEPRINT is where output from the PRINT commands will be sent. DCARPARM is how you tell CSP/AD about the original MSL concatenation. In Figure 14.32, "M=usrrwmsl" tells CSP that the ddname "usrrwmsl" will serve as the read-write MSL for processing. "ROMSL=(romsl1,romsl2)" tells CSP that the ddnames of two read-only MSLs will be used with "romsl1" being in front of "romsl2" in the concatenation order. The concatenation order may be altered dynamically by the MSL command as long as the ddnames are there to support it. You may use any ddnames that make you feel good. "CMDIN" tells CSP that the input commands will come from the EZECIN ddname. You may alter this by entering "CMDIN=myddnam" to redirect the input commands. Many other (less frequently used) options are possible for input to DCAPARM. See the Administration manual for your environment for details.

CHAPTER 14 EXERCISES

Questions

1. The act of preparing a CSP application for execution is called GENERATION. What is being generated?
2. During generation, what does the target environment signify?
3. Where does output from generation go?
4. How do you execute using the generated output?
5. When given a member name, how do CSP utilities decide which MSL's members to use?

6. What format is usually used to EXPORT/IMPORT from an MSL to an ALF and back into another MSL?
7. What format is used to EXPORT a member in a form that may be modified using normal system editors?
8. What symbol is used to delimit CSP/AD batch commands?
9. When executing CSP/AD in MVS batch, what ddname should contain the CSP/AD commands?
10. When executing CSP/AD in MVS batch, how does the system know which MSLs to use and their concatenation order?

Computer Exercise
Generating and Testing Applications

1. Generate the application used in the last chapter's exercise for execution in the same environment CSP/AD is running under.
2. Execute the generated transaction; see if it still works (it should!).
3. Print the application using the online facility. Be sure to include a cross-reference. If your system isn't set up for online printing, try doing it in batch.
4. If you haven't already done so, EXPORT the MS10A application and all its associates to the FZERSAM ALF from the UTILMSL. Then IMPORT it to your read-write MSL. (Hint: Follow the instructions earlier in this chapter!)
5. Print your application in batch.
6. Generate your application in batch.

External Source Format (ESF)

After that last chapter, this should be a breeze. External Source Format was introduced in the discussion on EXPORT and IMPORT utilities. Now we'll take a closer look.

Beginning with Version 3.2 of CSP, IBM has provided the External Source Format (ESF) definition facility. The External Source Format is designed to provide portability among AD/ CYCLE products. (AD/CYCLE is IBM's standard for application development.) Several currently available CASE (Computer Aided System Engineering) tools generate External Source Format during their construction phase. With the advent of Version 3.3, this facility has been improved considerably, principally by adding online capability where only batch could be used before.

One of the features of CSP/AD that is sometimes identified as a problem is that data is stored in VSAM KSDSs and we cannot edit them using ISPF/PDF and various other editing tools (yet). Another problem is that once a CSP application has been created, it is difficult to "clone" portions of it using the available CSP editing tools and utilities.

External Source Format (ESF) provides a partial solution. ESF allows the creation of 80-byte, sequential records that may be edited and processed using normal source code tools such as ISPF/PDF. After making changes to the ESF code, it may then be

IMPORTed back into CSP/AD. This facility both greatly aids making sweeping changes and the ability to "clone" (after all, isn't every great programmer a plagiarist?). "Cloning" is as simple as exporting an application, making global changes of application names and other names, then importing back into CSP/AD. Be careful when modifying External Source Format data. If you "mess up" the syntax, the data will not import successfully.

WHAT DOES ESF LOOK LIKE?

ESF has its own language and syntax. The basic elements are called "tags." Each type of member in an MSL is represented by a tag and associated keywords, values, and content lines.

Note the use of the period (.) as a start and end tag delimiter in Figure 15.1. The period is optional unless text follows a tag, but why not use it all the time to cut down on confusion? You can either key in your own code in ESF format (a daunting task) or EXPORT an application to ESF format and play with it (a much easier job).

```
:tagname
    keyword=value
    tag related text
:etagname.
```

Figure 15.1.   ESF syntax.

The :APPL . . . :EAPPL sequence in Figure 15.2 defines an application called XA01A. Notice that the main procedure and prologue are part of the application definition.

Figure 15.3 illustrates the basic definition of a process. Code shown here is exactly as it appears in the MSL. As long as CSP/AD syntax rules are followed, this code may be modified, added to, or deleted as necessary.

Maps and their fields are defined using the format shown in Figure 15.4. Records are defined in a similar fashion. Notice the :cfield . . . :ecfield sequence for each constant field defined on the

```
:EZEE 330          07/23/91 17:22:00
:appl    name    = XA01A    type    = MAIN
         date    = '07/23/91'
         time    = '10:27:07'
         mapgroup = XA01G
         pfequate = Y       implicit = N
:mainprc name    = XA01-MAINLINE.
:emainprc.
:prol.
Sample of CSP class, first application exercise

This program does nothing but display a map
:eprol.
:eappl.
```

Figure 15.2.   Sample ESF application specification.

map. Also notice the :vfield ... :evfield for the variable field XA01-PART. This sequence would be repeated for all variable fields on the map. The map fields are surrounded by :MAP and :EMAP tags. The ESF records may be modified as necessary using any editor that will work with an 80-byte fixed-length record (which means just about all editors). When creating table-style maps, it is often easier to define the first row, assigning attributes, edits, and names, then EXPORT the definition in ESF format. Next, an editing program (like ISPF/PDF) is used to duplicate the fields associated with the first row, changing position and name as necessary. Finally, the definition is IMPORTed back into CSP/AD. If sequential files are defined to the CSP/AD session, online use of the EXPORT and IMPORT utility makes this an easy task.

```
:process  name    = XA01-MAINLINE
          date   = '07/23/91'        time   = '10:35:53'
          option  = EXECUTE          refine  = N
          desc    = 'Mainline logic'.
:before.
PERFORM XA01-SENDMAP;            /* DISPLAY THE MAP
EZECLOS;                         /* EXIT THE APPLICATION
:ebefore.
:eprocess.
:process  name    = XA01-SENDMAP
          date   = '07/23/91'        time   = '10:35:53'
          option  = CONVERSE          object  = XA01M01
          refine  = N
          desc    = 'Map display routine'.
:before.
; /* THIS HAPPENS BEFORE THE MAP IS SENT
:ebefore.
:after.
; /* THIS HAPPENS AFTER THE MAP RETURNS
:eafter.
:eprocess.
```

Figure 15.3.   Sample ESF process specifications.

OKAY, SO HOW DO WE USE ESF?

With ESF you have four choices.

1. Create ESF data using an editor, and import it into CSP (perhaps the worst possible choice).
2. Export existing CSP members to ESF format, play with them, then import back into CSP as new/replacement members.
3. Create members using the PWS (Programmable Workstation), then export them to a file that is uploaded to the mainframe and then imported into CSP.

```
:map      grpname = XA01G    mapname = XA01M01
          date    = '07/23/91' time    = '10:38:33'
          mapsize = 024 080   startpos = 0001 0001
          devices = 3278-2B.
:present  varfold = N   deffold = N   spacer = '/'
          variable = '^'  constant = '#'
          tabpos  = 001.
:cfield   row = 001  column = 024
          type = CHA  bytes = 00034
.CSP CLASS - PART INVENTORY INQUIRY
:cattr    hilite  = USCORE   intense = NORMAL  protect = ASKIP
          color   = PINK     data    = ALPHA
          enter = N  mdt = N  fill = N  cursor = N  detect = N.
:ecfield.
:cfield   row = 001  column = 059
          type = CHA  bytes = 00363.
:cattr    hilite  = NOHILITE intense = NORMAL  protect = ASKIP
          color   = MONO     data    = ALPHA
          enter = N  mdt = N  fill = N  cursor = N  detect = N.
:ecfield.
:cfield   row = 006  column = 023
          type = CHA  bytes = 00019
.Enter PART ID ===
:cattr    hilite  = NOHILITE intense = NORMAL  protect = ASKIP
          color   = WHITE    data    = ALPHA
          enter = N  mdt = N  fill = N  cursor = N  detect = N.
:ecfield.
:vfield   row = 006  column = 043
          type = CHA  bytes = 00009
          name = XA01-PART
          desc = 'PART NUMBER              '
          editordr = 001.
:mapedits mininput = 04
          inputreq = Y.
:vattr    hilite  = RVIDEO   intense = NORMAL  protect = UNPROTECT
          color   = YELLOW   data    = ALPHA
          enter = N  mdt = N  fill = N  cursor = Y  detect = N.
:evfield.
```

Figure 15.4.   Sample ESF map definition.

```
:cfield    row = 006   column = 053
           type = CHA   bytes = 00216.
:cattr     hilite   = NOHILITE  intense = NORMAL  protect = ASKIP
           color    = MONO      data    = ALPHA
           enter = N  mdt = N  fill = N  cursor = N  detect = N.
:ecfield.
:cfield    row = 009   column = 030
           type = CHA   bytes = 00012
.Description:
:cattr     hilite   = NOHILITE  intense = NORMAL  protect = ASKIP
           color    = TURQUOISE data    = ALPHA
           enter = N  mdt = N  fill = N  cursor = N  detect = N.
:ecfield.
:vfield    row = 009   column = 043
           type = CHA   bytes = 00030
           name = XA01-DESCRIPTION
           desc = 'PART DESCRIPTION             '
           editordr = 002.
:vattr     hilite   = NOHILITE  intense = NORMAL  protect = ASKIP
           color    = MONO      data    = ALPHA
           enter = N  mdt = N  fill = N  cursor = N  detect = N.
:evfield.
:cfield    row = 009   column = 074
           type = CHA   bytes = 00115.
:cattr     hilite   = NOHILITE  intense = NORMAL  protect = ASKIP
           color    = MONO      data    = ALPHA
           enter = N  mdt = N  fill = N  cursor = N  detect = N.
:ecfield.
:cfield    row = 011   column = 030
           type = CHA   bytes = 00012
.Qty on Hand:
:cattr     hilite   = NOHILITE  intense = NORMAL  protect = ASKIP
           color    = TURQUOISE data    = ALPHA
           enter = N  mdt = N  fill = N  cursor = N  detect = N.
:ecfield.
```

Figure 15.4.   Sample ESF map definition (continued).

```
:vfield   row  = 011   column = 043
          type = CHA   bytes  = 00006
          name = XA01-QTY-ON-HAND
          desc = 'PART QUANTITY ON HAND        '
          editordr = 003.
:vattr    hilite  = NOHILITE  intense = NORMAL  protect  = ASKIP
          color   = MONO      data    = ALPHA
          enter = N  mdt = N  fill = N  cursor = N  detect = N.
:evfield.
:cfield   row  = 011   column = 050
          type = CHA   bytes  = 00138.
:cattr    hilite  = NOHILITE  intense = NORMAL  protect  = ASKIP
          color   = MONO      data    = ALPHA
          enter = N  mdt = N  fill = N  cursor = N  detect = N.
:ecfield.
:cfield   row  = 013   column = 029
          type = CHA   bytes  = 00013
.Qty on Order:
:cattr    hilite  = NOHILITE  intense = NORMAL  protect  = ASKIP
          color   = TURQUOISE data    = ALPHA
          enter = N  mdt = N  fill = N  cursor = N  detect = N.
:ecfield.
:vfield   row  = 013   column = 043
          type = CHA   bytes  = 00005
          name = XA01-QTY-ON-ORDER
          desc = 'PART QUANTITY ON ORDER       '
          editordr = 004.
:vattr    hilite  = NOHILITE  intense = NORMAL  protect  = ASKIP
          color   = MONO      data    = ALPHA
          enter = N  mdt = N  fill = N  cursor = N  detect = N.
:evfield.
:cfield   row  = 013   column = 049
          type = CHA   bytes  = 00831.
:cattr    hilite  = NOHILITE  intense = NORMAL  protect  = ASKIP
          color   = MONO      data    = ALPHA
          enter = N  mdt = N  fill = N  cursor = N  detect = N.
:ecfield.
```

Figure 15.4.   Sample ESF map definition (continued).

```
:vfield   row = 024   column = 001
          type = CHA   bytes = 00068
          name = EZEMSG
          editordr = 005.
:vattr    hilite  = NOHILITE  intense  = NORMAL  protect = ASKIP
          color   = RED        data     = ALPHA
          enter = N  mdt = N  fill = N  cursor = N  detect = N.
:evfield.
:cfield   row = 024   column = 070
          type = CHA   bytes = 00033.
:cattr    hilite  = NOHILITE  intense  = NORMAL  protect = ASKIP
          color   = MONO       data     = ALPHA
          enter = N  mdt = N  fill = N  cursor = N  detect = N.
:ecfield.
:emap.
```

Figure 15.4.   Sample ESF map definition (continued).

4. Export existing members to ESF format, download to a PWS, import to CSP/PWS, modify using the PWS, then export them to a file that is uploaded to the mainframe and then imported into CSP.

This is a good sequence of events to follow:

1. Decide which application will make a good model.
2. Export the application with all associates to a sequential file.
3. Use ISPF/PDF or some other editor to "play with the data."
4. Import the changed file into an empty MSL (just in case).
5. Import the member from the first MSL to your source MSL via your default ALF.
6. Take it for a test ride.

Before entering CSP/AD, it is a good idea to define sequential files that may be used for EXPORT and IMPORT duties. MVS users may even use members of PDSs (Partitioned Data Sets) for this purpose. I usually suggest having two files defined, one for EXPORTed output, the other for IMPORTing.

When using the CSP PWS (Programmable WorkStation), definitions are transported between the PWS and the host system using External Source Format. The next release of the PWS (scheduled for release late 1992) will include the "PWS EXPORT—UPLOAD—HOST IMPORT" and "HOST EXPORT—DOWNLOAD—PWS IMPORT" functions set up as single commands to simplify the transfer process.

ONLINE ESF EXPORT

Use CSP/AD's EXPORT utility either online or in batch to export selected MSL member(s) in External Source Format. Either mechanism works as long as the output file exists before EXPORTing begins and it is available to the CSP/AD session.

The EXPORT illustrated by Figure 15.5 is being used to export the application "XA01A" to a serial file identified with the

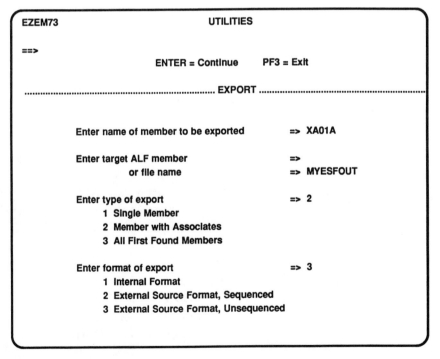

Figure 15.5.   ESF export.

file name (ddname) "MYESFOUT." The file name must be allocated so that your CSP/AD session can find it. Type of export tells CSP the scope of the export. Usually when exporting an application, it is desirable to specify "Member With Associates" to get not just the application definition but also the definition of its processes, maps, tables, and the data items used by them. If a member is in more than one MSL, only the first MSL's member is exported (first found).

Format of the export's 80-byte output will be either Sequenced or Unsequenced. Choose "Sequenced" to cause the External Source Format records to contain sequence numbers in columns 73–80. Choose Unsequenced to create the records without sequence numbers. The sequence numbers may seem like a good idea but actually tend to get in the way. Since you probably won't be punching cards (that could then be dropped) representing your External Source Format output, choose Unsequenced.

BATCH ESF EXPORT

Sometimes it is easier to do exports in batch mode, especially when multiple members will be exported at once. The output files must all be defined. If using a PDS (Partitioned Data Set) for output, each exported file may be placed in a different member.

Batch EXPORT (Figure 15.6) is useful when EXPORTing External Source Format for editing or transfer to the PWS (Programmable Workstation). MEMBER lists the name of the MSL member to be exported. Specify either SERIAL to name a sequential file's ddname/dlbl for output, or FILE to name an ALF's file name (ddname or DLBL). When specifying an ALF, you may also specify TARGET to choose the member name on the output ALF. Usually, sequential files are used to EXPORT and IMPORT External Source Format data.

```
EXPORT  MEMBER(XA01A)  SERIAL(MYESFOUT)
                TYPE(ASSOCIATES)
                FORMAT(EXTERNALU);
```

Figure 15.6. CSP/AD batch ESF export.

TYPE describes the scope of the EXPORT. MEMBER means that only the named member is to be exported. ASSOCIATES means that all members associated with the source member will be exported on a first-found basis (first occurrence in the MSL concatenation is exported). ALL should be used with caution; it means to export all members from all MSLs in the current MSL concatenation on a first-found basis. Format describes the future use of the EXPORTed member(s). EXTERNAL and EXTERNALU show that External Source Format is to be used. EXTERNAL creates 80-byte records with sequence numbers in columns 73–80 (in case you drop that deck). EXTERNALU creates 80-byte records with blanks in columns 73–80. It is usually simpler and cleaner to use EXTERNALU.

ONLINE ESF IMPORT

Use the CSP/AD IMPORT utility either online or in batch mode to bring External Source Format data into CSP/AD. CSP will detect that the incoming file is in External Source Format and will edit the syntax and definitions for correctness. Files containing the External Source Format definitions must be made available to CSP/AD. Each set of imported definitions should reside in a file by itself or by individual members of a PDS (Partitioned Data Set).

If the member being imported was EXPORTed (Figure 15.7) to an ALF previously in internal format, specify the ALF name, followed by a period (.), followed by the member name on the ALF. If the member being imported is in a sequential file, specify the file (ddname) containing the data. This file (ddname) must already be defined to your CSP/AD session. Specify REPLACE to copy over an existing member with the results of the import (make sure you have not made a mistake).

Selective import allows you to limit the members being imported. This is not normally done when importing External Source Format data. You may choose "On Error, Import Valid Members" when importing members stored in External Source Format. The Import utility validates the CSP syntax as it imports External Source Format code. If "On Error, Import Valid Members" is marked yes, Import will write individual members

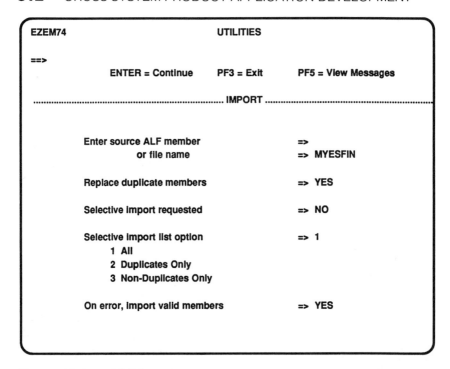

Figure 15.7. ESF import.

to the MSL as they are verified. If the field is marked no, Import must hold all members in memory until the last one passes (this will run much slower than specifying yes).

BATCH ESF IMPORT

Batch import is often desirable for processing imports. Batch is especially useful when importing many External Source Format files at once.

See Figure 15.8. To IMPORT from a sequential file (or PDS member in MVS), specify SERIAL and name the ddname/dlbl

---

IMPORT SERIAL(MYESFIN) REPLACE(Y) BYPASS(N)

Figure 15.8. CSP/AD batch ESF import.

where the EXPORTed data may be found. To import from an ALF, specify FILE and name the ddname/dlbl for the ALF; also specify SOURCE to identify the member in the ALF to be IMPORTed. It is most common to use sequential files when EXPORTing and IMPORTing External Source Format data. CSP will detect that the data being IMPORTed is in EXTERNAL format and edit it for syntax and definition errors. Specify BYPASS(Y) to skip invalid definitions and continue the IMPORT (the default). To cause the IMPORT to halt when an error is found, specify BYPASS(N).

If you want IMPORT to stop when an IMPORTed member has the same name as an existing member in the read-write MSL, specify REPLACE(N). Specify REPLACE(Y) to copy over any like-named members in the MSL. REPLACE(N) is the default.

DOCUMENTATION AVAILABLE ON ESF

SH20-6433	CSP/AD External Source Format Reference Version 3 Release 3
SH20-0520	CSP/AD External Source Format Reference Summary
GG24-3423	Cross System Product External Source Format Solutions

CHAPTER 15 EXERCISES

Questions

1. What type of file is usually output when exporting External Source Format?
2. What size records are used by External Source Format?
3. What is the primary benefit of sequence numbers?
4. What online option makes sure that an IMPORT does not accidentally overwrite existing MSL members?
5. What batch option makes sure that an IMPORT does not accidentally overwrite existing MSL members?
6. What online option is used to stop importing when a syntax or definition error is found in the External Source Format data being input?

7. What batch option is used to stop importing when a syntax or definition error is found in the External Source Format data being input?
8. What are the steps involved in moving data to a CSP PWS?
9. What are the steps involved in moving data back to the host from a CSP PWS?
10. What is the normal scope of data exported/imported using External Source Format?

Computer Exercise
Creating a "Cloned" Application using ESF

Choose one of the applications generated earlier for this exercise to be exported and cloned. The DB2 exercise makes for much more interesting reading!

1. Export your application with all associates to a sequential file.
2. Print out the ESF formatted file. (Use a standard print utility like IEBGENER.)
3. Make one or two global changes to the ESF file (maybe application and map names).
4. Import the ESF file to your MSL using a new name.
5. Test the application as it exists now.

CSP/AE ALF Utility

CSP provides a utility package specifically designed to help you manipulate ALFs and ALF members. The Application Load File Utility is officially called (ALFUTIL). ALFUTIL is usually stored in the ALF called UTILALF. (Cute!) The ALF utility is normally available in all CSP/AE environments. The command to enter the ALF utility is very similar from system to system, Figure 16.1 shows how to get in from CICS. Figure 16.2 shows how to get in from TSO.

In both of the above commands, and in all other environments, you simply execute CSP/AE (XSPE normally), then tell it which ALF to use (UTILALF) and which application to execute (ALFUTIL). If your installation has a panel-driven CSP/AE execution, or if the first screen you see prompts you for ALF name and application name, just enter UTILALF and ALFUTIL respectively. If the UTILALF is not available to you for some reason, call the CSP system administrator.

The first screen to display in the ALF utility is the "APPLICATION LOAD FILE UTILITY—ALF AND FUNCTION SPECIFICATION" panel shown in Figure 16.3. This is the "main menu" for the ALF utility. Choose a function and name the ALF(s) involved. The primary ALF is always INPUT to any ALF utility operations and is sometimes modified by the ALF utility. If un-

---

| **XSPE UTILALF.ALFUTIL** | **(CICS)** |

Figure 16.1. Executing the ALF utility in CICS.

| **XSPE APPL(UTILALF.ALFUTIL)** | **(TSO/ISPF)** |

Figure 16.2. Executing the ALF utility in TSO.

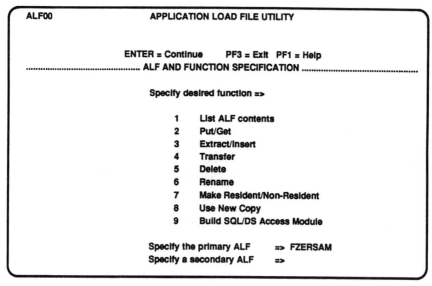

Figure 16.3. ALF utility.

specified, the ALF utility assumes the primary ALF is "FZERSAM." The secondary ALF is used only when doing the TRANSFER function. The secondary ALF is the OUTPUT for the TRANSFER function. Both ALFs (if used) must be allocated and available to the CSP/AE session or the transfer will fail. (And when it fails, it often does so by blowing you completely out of the ALF utility and CSP/AE!)

The different functions supported by the ALF utility are listed below:

Function	Description
List ALF Contents	Lists all members in the ALF. Abbreviated ALF commands may be used on this display (see Figure 16.5).
Put/Get	Used to copy ALF members to transportable files or to read an ALF member from transportable files.
Extract/Insert	Copy/read ALF member to/from an output file for transport between CICS, CMS, and TSO systems.
Transfer	Copy members from the primary ALF to the secondary ALF.
Delete	Delete ALF members.
Rename	Copy ALF member using new name.
Resident/Nonresident	(MVS only) Causes ALF members to be loaded only once rather than being loaded and unloaded as they are called.
Use New Copy	CICS only, allows access to latest version of resident ALF member. (Does not replace CEMT new copy for static load modules)
Build SQL/DS Module	SQL/DS under VSE only, helps optimize SQL/DS performance.

ALF CONTENTS

ALFs may contain a variety of members including exported MSL members, generated members, and test facility traces. The list below describes the various types of ALF members. I hope this is more information than you will ever need, but just in case you need it, here it is.

ALF Member Type	Description
Reference Files	Requested during generation, name is application name followed by dollar signs ($) and an "R" suffix.
Exported Members	Produced by the CSP/AD EXPORT utility from MSL source members. Names are specified during the EXPORT process.
Trace files	Created upon request when using the CSP/AD test facility. Names are specified at the time of the TEST.
Application Image Module	Generated application, first module executed by CSP/AE. Uses application name.
Processing Load Modules	Contains processing logic for the application including process statements, map edits, and literals. Named by appending the application name with dollar signs ($) and a letter "P" suffix.
Table Load Modules	Contain data table definitions for generated tables. Named by appending the table name with zeros (0) and a letter "T" suffix.
Map Load Modules	Generated map groups. Names consist of map group name suffixed by a two-character device type code.
Data Load Modules	Data structure and data item locations. Named by appending the application name with dollar signs ($) and a "D."
Data Characteristics Load Modules	Definitions of data items. Named by appending the application name with dollar signs ($) and a letter "I" suffix.

SQL Statement Preparation
Load Modules

Contains data needed to process SQL in an application. Named by padding the application name with dollar signs ($) and a letter "A" suffix.

SQL Statement Execution
Load Modules

Holds information needed by SQL processing in STATIC mode or using the EXECUTE IMMEDIATE interface. Named by appending the application name with dollar signs ($) and a letter "B" suffix.

DB2 Static Modules

Created upon request during generation. Contain Assembler statements that must be preprocessed and assembled into a "static load module" for SQL execution. Named by appending dollar signs ($) to the application name and adding an "X" suffix.

ANSI Static Modules

If used, contains Assembler statements that must be preprocessed and assembled into a "static load module" for SQL execution. Named by appending zeros (0) to the application name and adding an "X" suffix.

Data Description Specification
Load Modules

Optionally built to hold data description specifications for use in AS/400 systems. Named by appending the application name with dollar signs ($) and adding a "Q" suffix.

LIST ALF CONTENTS

This lists members in ALF using a tabular display (Figure 16.4). If requested, CSP will also list the number of records in each

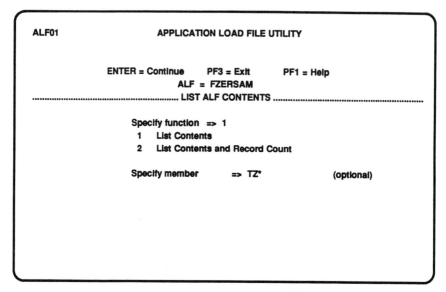

Figure 16.4. ALF contents selection.

member. If you do not want to list the entire ALF, use the "Specify member" field with an asterisk (*) wild card. Only fields matching the mask will be displayed. If nothing is entered in the "Specify member" field, all ALF members will be listed.

Each of the ALF utility options may be performed as line commands (in abbreviated form) from the list of ALF contents. See Figure 16.5. Place the abbreviated ALF utility commands as necessary (one at a time) next to selected members.

Abbrev. Command	Description
C	New copy (member)
CA	New copy (application)
D	Delete
DA	Delete application
E	Extract application
EM	Extract member
EX	Extract application and its associated maps and tables

G	Get member
N	Make member nonresident
NA	Make application nonresident
P	Print
PM	Put member to output file
R	Make member resident
RA	Make application resident
S	Pass member name to next ALFUTIL function selected
T	Transfer to secondary ALF
TA	Transfer application to secondary ALF
TX	Transfer application and its maps and tables to secondary ALF
V	View

All listings are written to the file (ddname) EZEP.

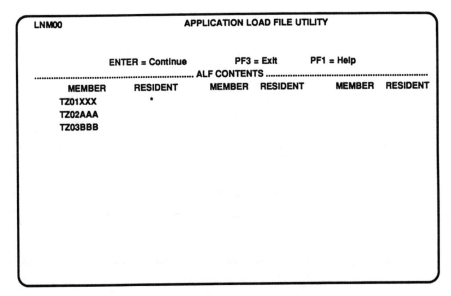

Figure 16.5.   ALF contents display.

PUT/GET

Put/Get are used to copy ALF members to transportable files or read ALF members from transportable files (Figure 16.6). Enter the function type. PUT writes SQL static modules and External Source Format members as 80-byte records to an output file named "ALFA" that must already exist and be allocated. Other members are written as 256-byte records to a file named "ALFP" (also must already exist and be allocated). GET copies members into the primary ALF from SQL static modules stored on a file, generated objects stored on a file, and exported objects stored on a file. External Source Format data to be input with GET must reside in 80-column records in a file named "ALFA." Other types of input files must be stored in 256-byte records using "ALFG" as the file name. Two additional options possible for GET: Specify "External Source Format" (if applicable), or "Other" when the input is from exported MSL data or generated modules.

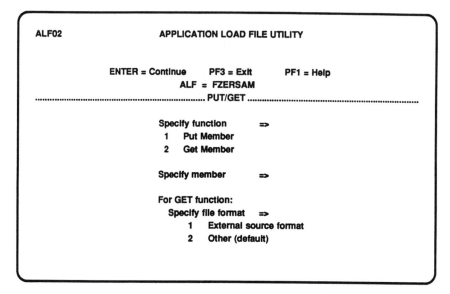

Figure 16.6. Put/Get.

EXTRACT/INSERT

Extract copies ALF members to a sequential file so that they may be transported between CICS, CMS, and TSO systems. The sequential file created may also be shared with AS/400 systems. Insert is the opposite, copying members from sequential files (from an ALF export) into the ALF. See Figure 16.7.

Specify the desired function. Be cautious about extracting or inserting entire ALFs. This could take a considerable amount of time (be sure to increase buffers if processing in batch). Insert member is able to select the desired member from the input file. When inserting, if duplicates are found, you are asked to make a decision: replace the existing member or cancel the insert. If you cancel after selecting "Insert Entire File," the insert process stops, but all members previously inserted remain on the ALF. When "Insert Member" is specified, only member type 3 may be specified. Member type describes the type of data to be written. If options 1 or 2 are selected (Generated Application), only one file is written.

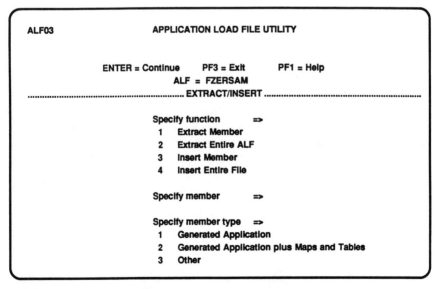

Figure 16.7. Extract/Insert.

```
┌─────────────────────────────────────────────────────────────────────┐
│                                                                       │
│   ALF04                    APPLICATION LOAD FILE UTILITY              │
│                                                                       │
│                                                                       │
│              ENTER = Continue     PF3 = Exit       PF1 = Help         │
│                            ALF = FZERSAM                              │
│       .................................... TRANSFER TO ............... │
│                                                                       │
│                 Specify member name  =>                               │
│                                                                       │
│                 Specify member type  =>                               │
│                                                                       │
│                      1  Generated Application                         │
│                      2  Generated Application plus Maps and Tables    │
│                      3  Other                                         │
│                                                                       │
│                                                                       │
│                                                                       │
│                                                                       │
│                                                                       │
└─────────────────────────────────────────────────────────────────────┘
```

Figure 16.8.   Transfer.

TRANSFER

Transfer is used to copy members from the primary ALF to the secondary ALF. This is the only command that uses the secondary ALF. See Figure 16.8.

Name the member to be transferred and indicate the type of member(s) to be transferred. Most often, this command is used to transfer "Generated Applications plus Maps and Tables" from a "test" ALF to a "staging" ALF, or from an ALF used in TSO to a CICS-defined ALF. Specifying a type of "Other" indicates that something other than generated members is to be transferred. When transferring to another ALF, you may have to take steps to make sure that the other ALF is available to you. For instance, an ALF described to CICS might not be available for the transfer until it is first closed in the CICS system.

DELETE

Delete is used to delete ALF members (Figure 16.9). The delete wipes out the named member immediately. No warning is given,

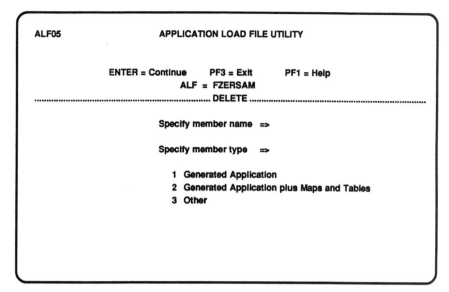

Figure 16.9.   Delete.

and there is no recovery (oops!). Be sure you really want to delete something first. When deleting applications, CSP does NOT delete associated maps and tables since other applications might be able to share them.

RENAME

Rename copies an ALF member using a new name (Figure 16.10). This is useful for backup purposes. The actual name of the CSP objects involved is not altered. Enter the name of the member to be changed, its new name, and the type of member it is. The ALF utility will not allow you to use a name that already exists on the ALF. After doing the rename, a panel appears that will allow deletion of the old member.

RESIDENT/NONRESIDENT

Modules are made resident to make them run faster (Figure 16.11). They run faster because they are loaded only once. For

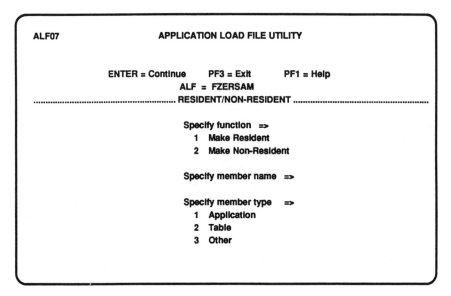

```
ALF06                    APPLICATION LOAD FILE UTILITY


              ENTER = Continue      PF3 = Exit        PF1 = Help
                           ALF = FZERSAM
.................................................. RENAME ..................................................

                    Specify member name  =>

                    Specify member type  =>

                        1  Generated Application
                        2  Other

                    Specify new name     =>
```

Figure 16.10.   Rename.

```
ALF07                    APPLICATION LOAD FILE UTILITY


              ENTER = Continue      PF3 = Exit        PF1 = Help
                           ALF = FZERSAM
.......................................... RESIDENT/NON-RESIDENT ..........................................

                    Specify function  =>
                        1   Make Resident
                        2   Make Non-Resident

                    Specify member name  =>

                    Specify member type  =>
                        1   Application
                        2   Table
                        3   Other
```

Figure 16.11.   Resident/Nonresident.

TSO users, this means they are loaded only once while running CSP/AE or CSP/AD. Under CICS, resident modules are loaded once and not unloaded as long as CICS remains active. By reducing the number of times the module(s) are loaded and unloaded, a considerable time savings can result. If a module no longer needs to be in memory, it may be made Nonresident. An application, table, or map group must be made resident with the ALF utility each time it is generated.

USE NEW COPY

This command (Figure 16.12) is used under CICS only. New copy replaces the current resident version of an application, table, or map group without waiting for CICS to shut down.

When resident applications, tables, and map groups are regenerated, two things should be done. First, the module should be made resident with the ALF utility. Second, the Use New Copy function should be performed to make sure that the latest version(s) of the modules are resident in memory. Do *not* confuse this with CICS "new copy" operations performed with CEMT or other tools.

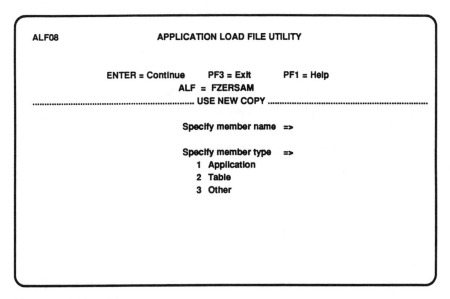

Figure 16.12.   New copy.

ALF UTILITY BATCH PROCESSING

The ALF utility may be run in batch mode. The MVS JCL that may be used is found in Figure 16.24, later in this chapter.

The CSP/AD product *may* be executed in batch mode in most environments. The *CSP/AD Developing Applications* manual documents the batch commands, while the individual *CSP/AD Administration* manual documents the process for each host environment. Most of the commands are identical in function to activities available online, but some commands are unique to the batch environment. Some illustrations in this chapter are for the MVS, but all commands work the same regardless of environment. To execute the CSP/AD commands in batch, the correct MSL and ALF file names must be used, and the syntax (shown in Figure 16.13) must be followed exactly.

Command syntax is unforgiving (as is most syntax) but basically pretty simple. Commands may not extend over multiple lines. Please note that the semicolon (;) must be used to delimit

- **Comments begin with /* and end with */ (don't put a slash in column 1)**

- **Commands may not be continued from line to line**

- **Parameters are keyword**

- **Case is unimportant**

- **Use a semicolon to delimit commands**

Figure 16.13. Batch CSP/AD syntax.

(end) each command. Illustrations of various batch commands follow. Commands work the same way in batch as they do interactively except for those commands that are peculiar to the batch environment. Batch execution of the ALF utility assumes that the name of the primary ALF is TRGTALF and that the name of the secondary ALF is SECNALF.

Put copies the named ALF member to either ALFA or ALFP (Figure 16.14).

```
PUT MEMBER(mbrnam);
```

Figure 16.14. Command to PUT an ALF member.

If ESF is specified, GET copies the member from the serial file ALFA into the ALF (Figure 16.15). If OTHER is specified (it is the default), GET looks for the input file in the ALFG file.

```
GET MEMBER(mbrnam) FORMAT(ESF or OTHER);
```

Figure 16.15. Command to GET an ALF member.

Extract is used to copy members from the primary ALF (TRGTALF) to a file named ALFO for transport to another CICS, CMS, or TSO system (Figure 16.16). Specify the member name and type (1 = generated application, 2 = generated application plus maps and tables, 3 = single member, 4 = table).

```
EXTRACT MEMBER(mbrnam) TYPE(1 or 2 or 3);
```

Figure 16.16. Command to EXTRACT an ALF member.

Insert copies from the ALFI input file into the primary ALF (TRGTALF) (Figure 16.17). If an entire file is to be inserted, omit the member name.

```
INSERT MEMBER(mbrnam);
```

Figure 16.17. Command to INSERT an ALF member.

Use Transfer to copy members from the primary ALF (TRGTALF) to the secondary ALF (SECNALF) (Figure 16.18). Member type should also be specified (1 = generated application, 2 = generated application plus maps and tables, 3 = single member).

TRANSFER MEMBER(mbrnam) TYPE(1 or 2 or 3 or 4);

Figure 16.18. Command to TRANSFER an ALF member.

Deletes specified member from the primary ALF (TRGTALF) (Figure 16.19). Member types 1 (generated application) and 3 (single member) may be deleted.

DELETE MEMBER(mbrnam) TYPE(1 or 3);

Figure 16.19. Command to DELETE an ALF member.

Copy member under a new name (Figure 16.20). Member types 1 (generated application) and 3 (single member) may be renamed.

RENAME MEMBER(mbrnam)
   TYPE(1 or 3)
   NEWNAME(newnam);

Figure 16.20. Command to RENAME an ALF member.

The command in Figure 16.21 makes a member resident. Member types 1 (generated application), 3 (single member), and 4 (table) may be made resident.

RESIDENT MEMBER(mbrnam) TYPE(1 or 3 or 4);

Figure 16.21. Command to make an ALF member RESIDENT.

The command in figure 16.22 makes a currently resident member nonresident. Member types 1 (generated application), 3 (single member), and 4 (table) may be made resident or nonresident.

```
NONRESIDENT MEMBER(mbrnam) TYPE(1 or 3 or 4);
```

Figure 16.22.   Command to make an ALF member NONRESIDENT.

The SET command (Figure 16-23) allows you to set a return code that will cause CSP/AE batch processing to stop. Enter a number between 0 and 16 or a STOPON keyword depending upon the severity you wish to allow. If a return code greater than that in the SET command occurs, processing will stop.

```
SET STOPON(retncode);
```

Figure 16.23.   Command to SET condition to stop CSP/AE.

Return Code	Usual Meaning
0	Everything is wonderful
4	CSP issued a warning
8	Error that may be survivable
12	Severe error
WARN	Same as 4
ERR	Same as 8

JCL FOR BATCH PROCESSING OF THE ALF UTILITY

Figure 16.24 shows some sample JCL for an MVS execution of the ALF utility in batch. Your CSP administrator probably has some JCL already made up that has been altered to fit your system's needs.

```
//xxx JOB
//stepnam  EXEC PGM=DCGBINIT
//**** CSP/AE LOAD LIBRARY
//STEPLIB  DD DSN=CSP322.AELOAD,DISP=SHR
//**** CSP MESSAGE FILE
//DCADZGF  DD DSN=CSP322.DZGMSG,DISP=SHR
//**** ALFO IS FOR EXTRACT OUTPUT
//ALFO     DD DSN=user.alfo,DISP=MOD
//**** ALFI IS FOR INPUT TO THE INSERT
//ALFI     DD DSN=user.alfi,DISP=SHR
//**** ALFA IS FOR PUT OUTPUT (80-BYTE)
//ALFA     DD DSN=user.alfa,DISP=MOD
//**** ALFP IS FOR PUT OUTPUT (256 BYTE)
//ALFP     DD DSN=user.alfp,DISP=MOD
//**** ALFG IS FOR GET INPUT
//ALFG     DD DSN=user.alfg,DISP=SHR
//**** UTILALF IS WHERE THE ALFBAT UTILITY IS STORED
//UTILALF  DD DSN=CSP322.UTILALF,DISP=SHR
//**** TRGTALF IS FOR THE PRIMARY, SECNALF IS FOR THE SECONDARY
//TRGTALF  DD DSN=user.primary.alf,DISP=SHR
//SECNALF  DD DSN=user.secndry.alf,DISP=SHR
//*
//EZEPRINT DD SYSOUT=*,DCB=(RECFM=VBA,LRECL=654,BLKSIZE=658)
//SYSPRINT DD SYSOUT=*
//**** ALF LOG FILE
//ALFL     DD SYSOUT=*,DCB=(RECFM=FBA,LRECL=121,BLKSIZE=1210)
//****
//**** EXECUTE APPLICATION ALFBAT, EZEUSER IS userid
//DCAPARM  DD *
A=UTILALF.ALFBAT U=userid
/*
//****
//**** ALFC IS WHERE BATCH COMMANDS GO FOR ALFBAT
//****
//ALFC     DD *
  /* BATCH ALF COMMANDS GO HERE */
  PUT MEMBER(XA01A$$X);
/*
//
```

Figure 16.24.   MVS JCL to execute ALF utility.

The CSP332 library prefixes will probably have to change for your installation; see your CSP administrator. The various input and output files should be allocated before executing this JCL. The TRGTALF ddname is required; it should point to the primary ALF to be used. SECNALF is required only for the TRANSFER command; it should point to the secondary ALF name.

In the DCAPARM statement, "A=UTILALF.ALFBAT" tells CSP/AE to execute the ALFBAT application found in the ALF named UTILALF. It also identifies the userID. The ALFC file contains any batch commands intended for the ALF utility. Be sure to follow syntax guidelines.

CHAPTER 16 EXERCISES

Questions

1. Where is the ALF utility executed?
2. What is the command to execute the ALF utility under TSO or CICS?
3. Which ALF is required for processing the ALF utility?
4. How can you see a list of ALF members?
5. How can you see a list of ALF members having names beginning with the characters "XA01"?
6. What ALF utility command is used as part of the DB2 STATIC LOAD MODULE processing?
7. When an ALF member that is an application is deleted, what is *not* deleted along with it?
8. Which command requires a secondary ALF?
9. Why are modules marked RESIDENT?
10. Should all modules be processed with USE NEW COPY after successful generation?

Computer Exercise
Moving Applications from ALF to ALF

1. Use the ALF utility to delete any members left over from CSP/ AD IMPORT and EXPORT exercises earlier.

2. Move your application from the ALF you generated into to another ALF you have access to (you may have to create one).
3. Test your application from the new ALF.
4. Under CSP/AD, TEST an application after directing the output to your ALF (FZERSAM probably).
5. Use the ALF utility to view, then delete the trace output.

17

TSO Considerations

Certain differences must be allowed for when executing CSP under its various environments. The Time Sharing Option (TSO) is the most popular environment for application development under MVS. TSO processing creates an address space and task for each user that is logged on. The overhead requirements of TSO make it a less attractive alternative for busy online production applications. Both CICS and IMS are designed for heavy throughput in an online situation and are better choices for busy online production applications.

CSP/AD is often executed under TSO because it is a favorite tool of existing application developers, and their prior skills may be used to enhance the development process. File locking and other considerations sometimes make it more difficult to run CSP/AD under CICS than under TSO. Most MVS installations run CSP/AD under TSO and CSP/AE under another environment (probably CICS or IMS). This is likely to change as the PWS (Programmable Workstation) is used more (and as it becomes more useable).

EXECUTING CSP/AD

In most installations, CSP/AD and CSP/AE will be set up as menu options from within the ISPF/PDF environment (Figure 17.1). When executing from within ISPF, IBM supplies an initial panel to "front end" CSP/AD. Every shop I have been in has either replaced this panel or modified it so don't be surprised if it doesn't appear in your installation.

Enter the ddnames of the read-write MSL and any read-only MSLs you will use in the spaces provided. This panel also establishes the beginning MSL concatenation sequence. If you specify a ddname that does not exist (you have not allocated yet), the "MSL SELECTION" panel will immediately display upon entering CSP/AD. Most installations make a major improvement to this panel by allowing you to enter the ddnames *and* dataset names of the MSLs, ALFs, and data files that will be used by the

```
                        CSP Application Development

      ===>


            R/W MSL   1     ===> mymsl

            R/O MSLs  2     ===> groupmsl    USERID        ===>
                      3     ===>            DBCS          ===>
                      4     ===>            DB2 Subsystem ===>
                      5     ===>            DL/I PSB      ===>
                      6     ===>            BKO           ===>
                                           BMP           ===>


            CMDIN         ===>              CMDOUT        ===>


            Invocation Parameter Group ===>


            Additional Parameters:
            ===> Function="TEST MEMBER(xa01a)"




      PF:  1/13=HELP        3/15=END
```

Figure 17.1. IBM-supplied CSP/AD invocation panel.

session. This is a good idea since the panel then performs the TSO allocations for you (behind the magical mirrors). Be careful! Many of these "home-grown" panels require that you deallocate on the way out of CSP/AD as well.

USERID is used to set the initial value for EZEUSR, if omitted. EZEUSR is set to the user ID being used. "DBCS => YES" tells CSP that double-byte character set (DBCS) data might be used. If using a terminal not capable of processing DBCS characters (not an IBM 5550 display or IBM PS/55) and DBCS data might be present, specify DBCS(YES) or "unpredictable results may occur." (Uh-oh!)

DB2 Subsystem tells CSP which DB2 subsystem to use. DL/I PSB, BKO, and BMP all have to do with IMS. PSB tells CSP which PSB may be executed. BKO tells CSP whether DL/I batch checkpoints are to be active. "BMP => YES" indicates that XSPD is to run as an IMS BMP job under TSO. Use "BMP => YES" when you need to test using IMS shared or fast path databases. CMDIN is used to list a file containing valid CSP/AD commands that are to be executed without viewing the panels. CMDOUT names the file where CSP should log commands, messages, and errors (fixed-length records, 121 bytes long, with ASA control characters in the first byte). Invocation Parameter Groups are explained in Chapter 18. The "Additional Parameters" line allows execution of single-function CSP/AD activities using the format: '=> Function="csp/ad function"'. See Figure 17.3 for a list of available functions and their format. CSP/AD may also be accessed by running the IBM-supplied command procedure XSPD. The syntax for running the XSPD clist is illustrated in Figure 17.2.

MSL tells CSP/AD the name of the read-write MSL. The ROMSL parameter names the read-only MSLs and the concatenation sequence. U (userID) tells CSP the initial value for EZEUSR, if omitted. EZEUSR is set to the user ID being used to execute XSPD. DSYS tells CSP which DB2 subsystem to use.

PSB, BKO, and BMP all have to do with IMS. PSB tells CSP which PSB may be executed. BKO tells CSP whether DL/I batch checkpoints are to be active. BMP(YES) indicates that XSPD is to run as an IMS BMP job under TSO. Use BMP(YES) when you need to test using IMS shared or fast path databases. DSYS and the IMS options may not be specified together.

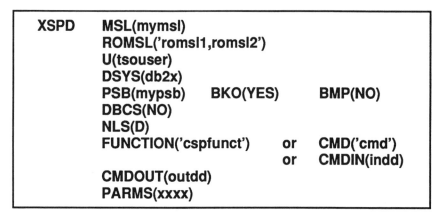

Figure 17.2. XSPD syntax (CSP/AD invocation).

DBCS(YES) tells CSP that double-byte character set (DBCS) data might be used. If using a terminal not capable of processing DBCS characters (not an IBM 5550 display or IBM PS/55) and DBCS data might be present, specify DBCS(YES) or "unpredictable results may occur." (Don't you love it?) NLS (National Language Support) is a single-character code indicating the language to be used for maps, messages, and tutorials. The possible values are "D" for mixed-case English (the default), "U" for upper-case English, "C" for simplified Chinese, "G" for German, "J" for Japanese, "K" for Korean, "P" for Portuguese, "S" for Spanish, "T" for traditional Chinese, and "W" for Swiss German. The language must have been installed in order to use this feature. The system administrator may change the default to fit your needs.

FUNCTION allows you to immediately enter a CSP/AD utility function in interactive mode. Figure 17.3 lists the XSPD functions. CMD and CMDIN are used to list a file containing valid CSP/AD commands that are to be executed without viewing the panels (also used for batch processing). CMDOUT names the file where CSP should log commands, messages, and errors (fixed-length records, 121 bytes long, with ASA control characters in the first byte). PARMS points to an Invocation Parameter Group (see Chapter 18).

```
COPY MEMBER(mbrnam);
DELETE MEMBER(mbrnam);
DIRECTORY MSL(mslnbr);
EDIT MEMBER(mbrnam)  TYPE(APPL or RECD or ITEM
                         or  TBLE or PSB or PROC
                         or  SGRP or MAP);
EXPORT MEMBER(mbrnam);
IMPORT SOURCE(alfmbr)   FILE(alfdd);
GENERATE MEMBER(mbrnam);
PRINT MEMBER(mbrnam);
RENAME MEMBER(mbrnam);
TEST MEMBER(mbrnam);
```

Figure 17.3.   XSPD function list.

EXECUTING CSP/AE

CSP/AE may be accessed by running the IBM-supplied command procedure XSPE. Care should be taken to make sure that the proper file allocations have been performed before executing either CSP/AD or CSP/AE. If running under ISPF/PDF, CSP provides a panel (that again is modified more often than not) to help at invocation time (Figure 17.4). This panel should be obtainable from a menu.

Enter the Application name to be executed in the space provided. The application must have been successfully generated for the TSO target environment in order to run it here. Tell CSP the name of the ALF where the generated modules can be found. The system default is the ALF with ddname FZERSAM. The ALF ddname specified (or FZERSAM) must be allocated before entering CSP/AE. If applicable, specify the desired National Language character: "D" for mixed-case English (the default), "U" for upper-case English, "C" for simplified Chinese, "G" for German, "J" for Japanese, "K" for Korean, "P" for Portuguese, "S" for Spanish, "T" for traditional Chinese, and "W" for Swiss German.

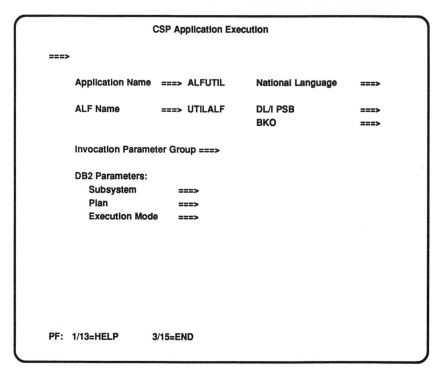

Figure 17.4. IBM-supplied CSP/AE invocation panel.

DL/I PSB and BKO are used only if IMS is involved. DL/I PSB tells CSP which PSB may be executed. BKO tells CSP whether DL/I batch checkpoints are to be active. The DB2 parameters should be filled in only when necessary. Subsystem tells CSP which DB2 subsystem to connect to. Plan names the DB2 application plan name to be used. If plan is omitted, the application name is used as the plan name. Execution mode describes what type of access is being performed: "D" if all access is using Dynamic SQL calls, "S" if all access is using static SQL (usually the preferred option), and "A" if all access is using ANSI static SQL. By specifying "D" you may override the generation options and execute in dynamic mode for this execution.

The syntax for running the XSPE clist is illustrated in Figure 17.5.

```
XSPE    APPL(alfnam.applnam)
        DSYS(db2x)     DMODE(x)      DPLAN(applnamx)
        PSB(mypsb)     BKO(YES)
        NLS(D)
        PARMS(xxxx)
```

Figure 17.5.   XSPE syntax (CSP/AE invocation).

The APPL parameter is used to tell CSP the name of the ALF and the name of the application to be executed. The ALF name goes first, followed by a period (.), followed by the application name (alfnam.applnam). If the ALF name and period are omitted, CSP assumes that the application is found in the ALF named "FZERSAM." DSYS tells CSP which DB2 subsystem to use. DMODE describes the type of SQL access to be used ("D" for Dynamic SQL calls, "S" for static SQL, and "A" to use ANSI static SQL). DPLAN names the application plan that corresponds to the application to be executed when using STATIC SQL. If DPLAN is omitted for static sql execution, the application name is assumed to be the application plan name as well.

PSB and BKO have to do with IMS. PSB tells CSP which PSB may be executed. BKO tells CSP whether DL/I batch checkpoints are to be active. DB2 and IMS options may not be specified together.

NLS (National Language Support) dictates the language to be used for maps, messages, and tutorials. PARMS points to an Invocation Parameter Group (see Chapter 18).

When executing CSP from ISPF, it is not usually a good idea to run ISPF from split-screen mode. You may find that things don't always work right when running CSP from a split screen. Besides, you cannot "toggle" between the sessions. Some users find it handy to create a CSP application that CALLs the ISPF dialog manager. This new application can be run from CSP/AD's test facility, allowing you to do ISPF/PDF things without exiting CSP.

FILE ALLOCATIONS

File allocations in the CSP/AD and CSP/AE environments are much simpler if command procedures or dialogs are created al-

ALLOCATE DD(myfile) DA('user.test.myfile') SHR REU

Figure 17.6.   TSO file allocation.

lowing user specification of the necessary files and ddnames. Otherwise, all users must become conversant with the TSO AL-LOCATE command and its use.

The TSO ALLOCATE statement (Figure 17.6) is the online equivalent to a DD statement in JCL. Enter the ALLOCATE from TSO READY mode or ISPF/PDF Option 6. You may also enter the command from any ISPF/PDF command line by using TSO as a prefix "TSO ALLOCATE . . .". If at any time you forget the syntax, simply enter "TSO HELP ALLOCATE" from any ISPF/PDF command line.

DD (DDNAME or FILE) tells TSO the ddname of the file to be used. This is the file name known to CSP/AD or CSP/AE. Be sure to pick a name that will be easy for you to remember. A common practice in CSP is to adopt a naming convention of having the last node of the DSN be the ddname. Some kind of naming convention is a good idea; otherwise you find yourself scrambling around looking for names. DA (DATASET or DSNAME) contains the dataset name of the file being allocated. If your TSO profile has PREFIX turned on (and most profiles do), your TSO userID is automatically added to the front of any dataset name specified without quotes (test.alf). Placing quote marks around a dataset name causes TSO to use the fully qualified name exactly as you enter it ('sysq.test.alf'). All of the options normally available on a JCL DD statement may be used with ALLOCATE. Fortunately we only use one or two besides DD and DA. Disposition of the dataset must be defined and should be set to SHR (share); other-wise TSO defaults to OLD. This means that no one else will be able to access the dataset until you leave TSO or issue a FREE command. Finally, REUSE allows you to ALLOCATE a DD name that is already in use and reuse it. Be careful, this is not a concatenation, it is a replacement for the old allocation. When you are done using a file, it is a good idea to FREE it so that others may allocate it. The TSO FREE command is used for this, or your allocations remain in force until you log off of TSO.

Enter FREE (UNALLOC) from TSO READY mode (Figure 17.7) or ISPF/PDF Option 6. You may also enter the command

FREE DD(myfile)

Figure 17.7.   TSO file unallocation (free).

from any ISPF/PDF command line by using TSO as a prefix "TSO FREE . . .". If at any time you forget the syntax, simply enter "TSO HELP FREE" from any ISPF/PDF command line. You may specify either DD (DDNAME or FILE) or DA (DATASET or DSNAME) on the FREE statement. *Do not use* the "ALL" option on the FREE statement if you intend to do anything at all with ISPF again in the same session. Your user profile causes many (sometimes many, many) allocations that you are normally unaware of. To get some idea of what's being done for you all the time, key in the following command from an ISPF/PDF command line: "TSO LISTALC ST." The "TSO LISTALC ST" command will list all ddnames and dataset names currently allocated to your session. It's normally quite a list.

It is often ungainly to continually execute ALLOCATE and FREE statements while testing your CSP applications. If you are lucky, your installation has supplied "front end" panels for CSP/AD and CSP/AE to automate this function. Another alternative is to create one CLIST or REXX EXEC to do the allocations and a second to perform the FREEing of datasets.

Figure 17.8 illustrates a REXX EXEC used to allocate several files. Your dataset and ddnames will probably be different, but this should give you an idea of how it works. Be careful! The comment (marked by /* and */) in the first line of the EXEC is necessary so that TSO will know it is dealing with a REXX EXEC rather than a CLIST. Let's pretend that REXX EXEC is stored in a partitioned dataset named "little.puppies" in a member called "barklots." The command to execute the REXX EXEC is illustrated in Figure 17.9.

Another way to execute the REXX EXEC is to use ISPF/PDF option 3.4 (dataset list) to list the partitioned dataset. Then, use option "M" next to the dataset name to create a member list (you've probably done this with "E" for edit or "B" for browse in the past). This will give you a list of members in the library. Key in option "EX" to execute the REXX EXEC (this also works with CLISTS). By the way, you may edit or browse members from the

```
/* THIS IS A REXX EXEC TO ALLOCATE FILES                     */
/* NEEDED TO RUN CSP TEST APPLICATIONS                       */
/*                                                           */
/* PLEASE FEEL FREE TO ADD/CHANGE/DELETE AS YOU SEE FIT      */
/*                                                           */
SAY 'HERE WE GO...'
"ALLOCATE DD(JJKESF)   DA('GROUPX.TEST.ESF')  SHR"
"ALLOCATE DD(TZINPUT)  DA('GROUPX.TEST.TZINPUT') SHR"
"ALLOCATE DD(CLSMSL)   DA('GROUPX.TEST.MSL')  SHR"
"ALLOCATE DD(CLSALF)   DA('GROUPX.TEST.ALF')  SHR"
"ALLOCATE DD(MS10F01)  DA('GROUPX.TEST.MSGFILE') SHR"
"ALLOCATE DD(DCACLSD)  DA('GROUPX.TEST.MSGFILE') SHR"
SAY 'ALL DONE NOW...'
```

Figure 17.8. TSO REXX EXEC to allocate files.

```
TSO EXEC 'little.puppies(barklots)'
```

Figure 17.9. Executing the REXX EXEC.

member list by using the "E" or "B" options (respectively) in front
of the member names.

Figure 17.10 shows the REXX EXEC to FREE the datasets
allocated by the previous REXX EXEC. Execute this new REXX
EXEC in exactly the same way as the first one. If you will be

```
/* THIS IS A REXX EXEC TO FREE FILES                         */
/* NEEDED TO RUN CSP TEST APPLICATIONS                       */
/*                                                           */
/* PLEASE FEEL FREE TO ADD/CHANGE/DELETE AS YOU SEE FIT      */
/*                                                           */
SAY 'HERE WE GO...'
"FREE DD(TZINPUT)"
"FREE DD(CLSMSL)"
"FREE DD(DCACLSD)"
"FREE DD(MS10F01)"
SAY 'ALL DONE NOW...'
```

Figure 17.10. TSO REXX EXEC to free files.

leaving TSO shortly after exiting CSP, leave this step out since all of your files will be FREEd automatically when your TSO session ends.

TESTING CICS AND SQL APPLICATIONS

When testing CICS and/or SQL applications under TSO, there are some things you need to do to make sure that the test environment adequately simulates the production environment. Most CICS applications will be run in "segmented" mode (see Chapter 18). Segmented mode causes the CICS task to end after each map is displayed to a user, beginning a new CICS task when the user responds (that's right, in the middle of the CONVERSE). The end of the CICS tasks forces a CICS SYNCPOINT that in turn causes a DB2 COMMIT to occur. Under the test environment, CSP/AD does not segment so this implicit SYNCPOINT/COMMIT does not occur. To make the test facility BEHAVE in the same fashion as the production environment, move a 1 to the special CSP field named EZECNVCM.

It is also possible for a program to take a segmented application and make it unsegmented (usually undesirable) by playing with a special field named EZESEGM. It is a good idea to move 1 to EZESEGM before every CONVERSE. If the value in EZESEGM is somehow changed to 0 before a CONVERSE, the application becomes unsegmented no matter what generation and execution options have been specified. SQL statements often return status codes (SQLCODEs) that indicate some warning or error condition. For reasons unknown to the author, CSP treats a non-zero response from SQL as a hardware error. These "hardware errors" will cause the application to end immediately UNLESS you move a 1 to the special CSP field named EZEFEC before any SQL statements occur in the application. Moving a 1 to EZEFEC tells CSP to not abort when a "hardware error" occurs, allowing the application to test the SQLCODE (using EZESQCOD) and decide what to do next.

To recap the extra moves that will help make testing CICS-DB2 programs under TSO test correctly:

Move 1 to EZECNVCM to mimic segmented commits.
Move 1 to EZESEGM prior to CONVERSEs to ensure segmentation.

Move 1 to EZEFEC to let the CSP application trap SQL errors.

Leave these statements in place when testing is complete. They cause no performance degradation at execution time.

COMMON LIBRARIES

CSP/AD and CSP/AE come with many programs and libraries that are used "behind the scenes" and are usually "invisible" to us. To store many of these for our use, CSP uses various MSLs and ALFs. Two libraries that are commonly available are UTILMSL and UTILALF. These libraries are allocated as part of CSP startup. UTILMSL is a utility MSL full of useful examples. The MS10A application and its components are stored in UTILMSL. UTILMSL is normally available any time you are in CSP/AD. UTILALF contains the members necessary to run the ALF utility. UTILALF is normally available any time you are in CSP/AE.

CHAPTER 17 EXERCISES

Questions

1. What is the TSO command to execute CSP/AD?
2. When using the above command, how can you tell CSP/AD which MSLs to use?
3. What is the TSO command to execute CSP/AE?
4. Using the above command, how would you cause CSP/AE to execute the application 'JOB' from the ALF named 'GOOD'?
5. What TSO command is used to make files available to a session?
6. How would you code the command to make the file with dataset name 'user.test.file' available to your TSO session using the ddname 'myfile'?
7. How would you code the commands to make the file with dataset name 'user.message.rrds' available to the TSO session using the ddnames 'MS10F01' and 'DCAXXXD'?
8. Why is the SHR option important?
9. What TSO command is used to release files when done?

10. What three moves should be added to applications when testing CICS-DB2 applications?

Computer Exercise
Create REXX EXEC to allocate files

1. Create the REXX EXEC to allocate the following files:

DATASET NAME	DDNAME
'test.inventory.books'	'books'
'user.message.file'	'dcazxzd'
'user.message.file'	'ms10f01'
'my.rw.msl'	'mymsl'
'your.ro.msl'	'yourmsl'
'my.test.alf'	'myalf'
'user.esf.file(output)'	'esfout'
'user.esf.file(input)'	'esfin'

2. Run your exec.
3. Use the command 'TSO LISTALC ST' from an ISPF/PDF command line to see the files currently allocated to your ID.
4. Create a REXX EXEC to free all files allocated previously and run it.
5. Use 'TSO LISTALC ST' to make sure they have been freed.

18

CICS Considerations

Both CSP/AD and CSP/AE function under CICS. Due to locking and buffer concerns, CSP/AD under CICS works best if every individual has his or her own MSL, message file, and ALF for testing purposes. Sometimes under CICS, lockout can occur when using common MSLs or test files. Having separate MSLs and test files for each user alleviates this problem.

To execute CSP/AD under CICS you use the XSPD transaction as illustrated in Figure 18.1. The MSL name (mslnam) is a positional parameter telling CSP/AD the name of the read-write MSL. The ROMSL parameter names the read-only MSLs and the concatenation sequence. If both MSL name and the ROMSL parameter are missing, CSP/AD will display the MSL SELECTION panel on startup.

"P=qnam" tells CSP the name of the transient data queue to send printed output to; the default is EZEP. RT names the CICS transaction to be executed after CSP/AD finishes. This is useful when attempting to keep the user in a menu-driven environment.

NLS (National Language Support) is a single-character code indicating the language to be used for maps, messages, and tutorials. The possible values are "D" for mixed-case English (the

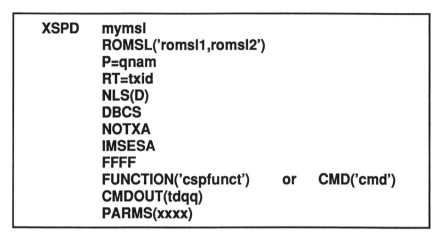

```
XSPD    mymsl
        ROMSL('romsl1,romsl2')
        P=qnam
        RT=txid
        NLS(D)
        DBCS
        NOTXA
        IMSESA
        FFFF
        FUNCTION('cspfunct')        or        CMD('cmd')
        CMDOUT(tdqq)
        PARMS(xxxx)
```

Figure 18.1.   XSPD syntax (CSP/AD invocation).

default), "U" for upper-case English, "C" for simplified Chinese, "G" for German, "J" for Japanese, "K" for Korean, "P" for Portuguese, "S" for Spanish, "T" for traditional Chinese, and "W" for Swiss German. The language must have been installed in order to use this feature. The system administrator may change the default to fit your needs.

DBCS tells CSP that double byte character set (DBCS) data might be used. If using a terminal not capable of processing DBCS characters (not an IBM 5550 display or IBM PS/55) and DBCS data might be present, specify DBCS or "unpredictable results may occur." (So watch out!)

NOTXA tells CSP that some CALLs to non-31-bit programs may be executed. CSP then acquires all user storage below the so-called "16-meg" line guaranteeing 24-bit addresses. (See the "24-bit versus 31-bit" section later in this chapter for more on this topic.) IMSESA is used to allow applications to use CSPTDLI to access the IMS/ESA message queue.

The FFFF parameter causes all CALLs from applications to non-CSP programs to include the length of the end-of-list indicator in the COMMAREA length and that the end-of-list indicator (four bytes with value x'FFFFFFFF') is passed along with other CALL parameters.

COPY MEMBER(mbrnam);

DELETE MEMBER(mbrnam);

DIRECTORY MSL(mslnbr);

**EDIT MEMBER(mbrnam) TYPE(APPL or RECD or ITEM
                          or TBLE or PSB or PROC
                          or SGRP or MAP);**

EXPORT MEMBER(mbrnam);

IMPORT SOURCE(alfmbr)   FILE(alfdd);

GENERATE MEMBER(mbrnam);

PRINT MEMBER(mbrnam);

RENAME MEMBER(mbrnam);

TEST MEMBER(mbrnam);

Figure 18.2.   XSPD function list.

FUNCTION allows you to immediately enter a CSP/AD utility function in interactive mode. Figure 18.2 lists the XSPD functions.

CMD must be a valid batch command enclosed in double quotes ("). CMDOUT names the transient data queue where CSP should log commands, messages, and errors (fixed-length records, 121 bytes long, with ASA control characters in the first byte). PARMS point to an Invocation Parameter Group (discussed later in this chapter).

To execute CSP/AE under CICS directly, you use the XSPE transaction code (Figure 18.3). Invocation parameter groups (discussed later in this chapter) allow the execution of CSP/AE using different transaction code. The first parameter is positional and tells CSP the name of the ALF and the name of the

```
XSPE    alfnam.applnam
        SEG
        P=qnam
        RT=txid
        DMODE=s
        NOTXA
        TSMS
        IMSESA
        FFFF
        PARMS(xxxx)
```

Figure 18.3.   XSPE syntax (CSP/AE invocation).

application to be executed. The ALF name goes first, followed by a period (.), followed by the application name (alfnam.applnam). If the ALF name and period are omitted, CSP assumes that the application is found in the ALF named "FZERSAM." The SEG parameter is necessary only if the application was generated to run in EITHER segmented or unsegmented mode. If the application was generated for EITHER, SEG causes the application to run in segmented mode.

"P=qnam" tells CSP the name of the transient data queue to send printed output to; the default is EZEP. RT names the CICS transaction to be executed after CSP/AD finishes. This is useful when attempting to keep the user in a menu driven environment. DMODE controls the execution mode for DB2 calls. "D" tells CSP to perform dynamic SQL despite generation parameters. "S" and "A" tell CSP to look for static SQL or ANSI static SQL modules and execute in static mode if possible. NOTXA tells CSP that some CALLs to non-31-bit programs may be executed. CSP then acquires all user storage below the so-called "16-meg" line guaranteeing 24-bit addresses. (See the "24-bit vs 31-bit" section later in this chapter for more on this topic.)

CSP stores all records, maps, and other data in CICS Temporary Storage between segmented (pseudoconversational) iterations. Before each CONVERSE the data is written to Temporary Storage. After the user responds, the data is written back into memory from Temporary Storage. CSP normally uses CICS Auxiliary Temporary Storage (stored on disk). Using the TSMS option causes CSP to use Main Temporary Storage instead (in virtual memory). Main Temporary Storage reads and writes are significantly faster than Auxiliary Temporary Storage. However, if your installation does not have a surplus of virtual memory the performance benefit will be negated and perhaps become a problem. (If you're not in an XA or ESA shop don't even think about it!)

IMSESA is used to allow applications to use CSPTDLI to access the IMS/ESA message queue. The FFFF parameter causes all CALLs from applications to non-CSP programs to include the length of the end-of-list indicator in the COMMAREA length, and that the end-of-list indicator (four bytes with value x'FFFFFFFF') is passed along with other CALL parameters. NLS (National Language Support) dictates the language to be used for maps, messages, and tutorials. PARMS points to an Invocation Parameter Group (discussed later in this chapter).

CICS TABLE ENTRIES

CICS processing is controlled by a series of tables. CICS's system tables define all resources available to CICS and a great deal of other information including limits on the resource use. Using CSP under CICS requires entries in various CICS control tables:

DCT	Destination Control Table
FCT	File Control Table
PCT	Program Control Table
PPT	Processing Program Table
RCT	Resource Control Table

Running CSP/AD under CICS requires the following entries (at least):

FCT	Entries for MSLs, ALFs, message files, and any data files
PPT	Entries for CSP/AD routines
PCT	Entry for XSPD

RCT   Entry for XSPD's plan

Using CSP/AE under CICS requires:

FCT   Entries for the ALFs, message files, and any data files

PCT   Entries pointing to DCBINIT for all main applications that will be referenced directly by a transaction ID

PCT   Entries pointing to DCBRINIT for those secondary main applications that will only be executed as a result of changing EZESEGTR prior to a CONVERSE PROCESS

PPT   Entries for DCBINIT, DCBRINIT, and any static load modules for applications making DB2 calls (usually the application name suffixed by an 'X')

RCT   Entries for any transaction IDs that will use DB2, pointing to the appropriate plan(s)

DCT   Entries for any Transient Data Queues to be used as sequential records by CSP applications

INVOCATION PARAMETER GROUPS

Version 3.2 of CSP introduced the capability to use INVOCATION PARAMETER GROUPS, also known as PARAMETER ENTRY GROUPS. In earlier versions of CSP, if you desired to use a transaction ID other than XSPE to execute a CSP application, a complex and high-maintenance procedure was followed:

1. A PCT entry for the application name is made.
2. The DCBINIT module is link-edited together with the application.

Version 3.2 began the use of Invocation Parameter Groups. Parameter Groups are stored in a VSAM KSDS usually called DCAPRMG. Next all PCT entries for main transaction IDs are set to point to the DCBINIT program. When a transaction OTHER than XSPE is used to invoke DCBINIT, the program looks in the Parameter Group file to find a record matching the transaction ID that was used. The Parameter Group record tells CSP/AE which application to execute and several other things.

XSPE PRMALF.PRMUTIL

Figure 18.4.   Running the parameter group utility.

Invocation Parameter Groups are created either in batch or online using a utility provided by CSP (Figure 18.4). With XSPE, execute the Parameter Group utility (PRMUTIL) that is normally stored in the ALF named (PRMALF). Check with your system administrator to make sure that DCAPRMG and PRMALF are available before trying this.

The first panel displayed by the Parameter Group Utility is shown in Figure 18.5. Enter the CICS transaction code to be used to execute a CSP application as the Parameter Group name. If the Parameter Group already exists, you may modify it. If the Parameter Group does not exist, you will be allowed to create one.

If you are unsure of the Parameter Group name you want to

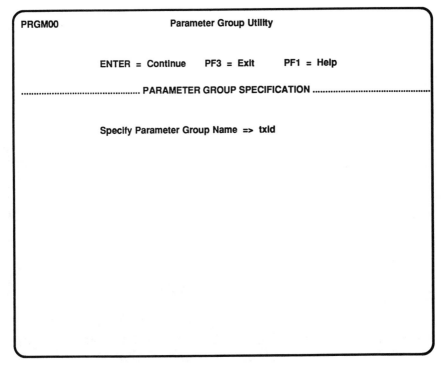

Figure 18.5.   Parameter group specification.

use, enter a question mark (?) as the Parameter Group name. This will cause the "PARAMETER GROUP LIST DISPLAY" panel to appear (Figure 18.6). Place an "S" before a Parameter Group to be modified. Place a "D" before a Parameter Group to be deleted. Be careful, the delete does not ask for verification, it assumes you meant to delete.

The PARAMETER GROUP DEFINITION panel (Figure 18.7) is used to place Parameter Group entries into DCAPRMG. While it is possible to execute CSP/AD with Parameter Entry groups,

```
PRGM02                    Parameter Group Utility


         ENTER = Continue    PF3 = Exit    PF4 = Refresh    PF1 = Help
           PF7 = Scroll Forward    PF8 = Scroll Backward

............................... PARAMETER GROUP LIST DISPLAY ...............................


__ tx01          __ tx02          __ tx03          __ tx04
```

Figure 18.6.   Parameter group list display.

they are almost exclusively used to execute XSPE in the CICS environment.

Parameter Group entries are specified using a simple, keyword-oriented syntax (Figure 18.8). "A=alfnam.applnam" is used to tell CSP which application to execute and what ALF its modules may be found in. The ALF name (and trailing period) may be omit-

```
┌─────────────────────────────────────────────────────────────┐
│  PRGM03                    Parameter Group Utility            │
│                                                               │
│                                                               │
│            PF3 = File and Exit    PA2 = Cancel   PF1 = Help    │
│                                                               │
│  ............................ PARAMETER GROUP DEFINITION ...........................  │
│                                                               │
│                                                               │
│      Parameter Group:                                         │
│                                                               │
│          =>                                                   │
│                                                               │
│                                                               │
│                                                               │
│                                                               │
│                                                               │
│                                                               │
│                                                               │
│                                                               │
│                                                               │
└─────────────────────────────────────────────────────────────┘
```

Figure 18.7.   Parameter group definition—Initial display.

ted. "A=applnam" and CSP will assume that the needed modules
are stored in the ALF called FZERSAM. SEG is used when segmentation is desired and the application is generated for EITHER
segmented or unsegmented processing. "P=tdqq" names the Transient Data queue to which print is directed. EZEP is the default
Transient Data queue for printed output. NLS specifies a language code. The default is D for mixed-case English. SMODE describes the type of SQL/DS access to be used. A=ANSI Static,
S=Static Module, and D=Dynamic execution. DMODE describes
the type of DB2 access to be used. A=ANSI Static, S=Static Module, and D=Dynamic execution. RT names the transaction to be
executed when CSP is done. TSMS causes CICS to store segmentation data in main temporary storage rather than auxiliary temporary storage. (This may require large quantities of virtual space.
In XA and ESA shops the speed gained is often worth it.) NOTXA
forces data to be stored below the 16M line. FFFF adds an end-of-

```
A=alfnam.applnam
SEG
P=tdqq
NLS=D
SMODE(A or D or S)   or   DMODE(A or D or S)
RT=txid
TSMS
NOTXA
FFFF
```

Figure 18.8.   Parameter group entry syntax options.

list indicator (x'FFFFFFFF') to COMMAREA when CSP applications CALL non-CSP programs.

Figure 18.9 shows the PARAMETER GROUP DEFINITION with a sample Parameter Group entry.

```
PRGM03                      Parameter Group Utility


           PF3 = File and Exit    PA2 = Cancel    PF1 = Help

...................................... PARAMETER GROUP DEFINITION ......................................


   Parameter Group:

       => A=alfnam.applnam P=EZEP DMODE=D TSMS
```

Figure 18.9.   Sample parameter group definition.

SEGMENTED APPLICATIONS
(PSEUDOCONVERSATIONAL TRANSACTIONS)

It has been generally accepted wisdom for years in CICS to generate pseudoconversational code. Pseudoconversational code usually improves CICS system throughput by having transactions free up system resources while waiting for user input. To accomplish this, transactions (usually) send maps to a screen, store data somewhere inside of CICS, and end telling CICS what transaction to execute when the user responds. When the user responds, CICS starts up the predetermined transaction and the stored data is retrieved for processing. This entails considerable programming effort in third-generation languages (3GLs) like COBOL and PL/I. After sending a map to the screen, an application has two choices:

1. Wait for the user to respond.
2. End telling the system how to restart things when the user responds.

The first practice, waiting upon a response, is called conversational programming. While the application waits, the storage allocated to it is kept, limiting the number of concurrent users possible. If an application has locked resources, they remain locked while the system waits. This sometimes causes contention problems because you don't know when the user is coming back.

Telling the system to restart when the user responds is called pseudoconversational programming. With this method, the program saves the storage associated with the application, freeing the main memory at the expense of the I/Os required to store the data. This allows greater concurrent use of the system since system resources are *implicitly* freed by the end of application segments (iterations). So, it doesn't matter when the user comes back.

CSP supports pseudoconversational processing with the segmentation feature. Segmentation is an automatic process invoked by CSP and requires no additional processing logic in the application. There are two techniques used to ensure that the application behaves the same way when tested under TSO as it does under CICS, and to allow proper Application Plan switching in DB2.

How Does Segmentation Work?

Segmentation is turned "on" as part of the generation process. Applications may be generated for Segmented, Unsegmented, or Either. If Either is specified, the SEG parameter is used at execution time (usually from a Parameter Entry Group) to control segmentation. *Caution!* Moving 1 to EZESEGM before a CONVERSE turns segmentation "off."

1. Each CONVERSE delineates a task.
2. A SYNCPOINT is implicit at the end of task caused by a CONVERSE, so any record locks are freed and database changes are committed (also any sequential operations are CLOSEd).
3. To cause another transaction to be executed upon reentry (and you need to switch DB2 Application Plans), place the new transaction code into EZESEGTR before issuing the CONVERSE. When using EZESEGTR as above, it is best to do all DB2 I/O after the CONVERSE. This makes sure that the correct Application Plan has been loaded by DB2 (the plan is switched by DB2 when the new task begins).
4. Move 1 to EZESEGM before CONVERSE executions to make sure that segmentation is not subverted.
5. Move 1 to EZECNVCM to test segmented applications under TSO with locking and commits similar to the CICS environment.

CONVERSE processes are divided into two portions: what happens *before* the CONVERSE and what happens *after* the CONVERSE (Figure 18.10). When Segmentation is turned on, execution halts while waiting for the user to respond. In CICS (and DPPX) systems this means that an implicit SYNCPOINT is caused by each CONVERSE.

Segmentation is only required/beneficial in CICS and DPPX systems since they are the only CSP supporting systems in which applications share an address space. Greater concurrent use of the system is possible when programs are segmented because the segmentation process frees memory for use by other applications (more applications may use the same amount of memory). CSP segmentation is achieved by saving all of the storage areas asso-

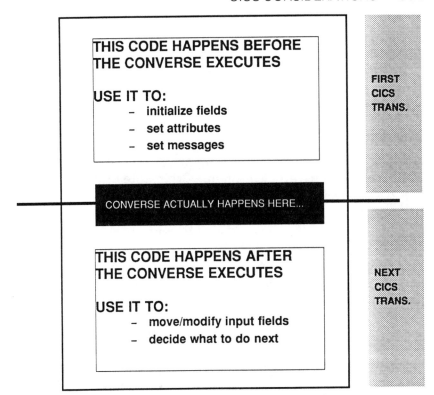

Figure 18.10. Converse segmentation.

ciated with an application to a CICS Temporary Storage queue. The I/O associated with writing the segment storage areas can be considerable, and if you have a small or very infrequently used (online) system, it may actually be more expensive than not segmenting. (I know, that sounds like CICS blasphemy! But recent improvements in hardware and CICS make conversational programming less "evil" every year.)

In most larger systems, the number of applications needing to run concurrently (yours and everybody else's) makes segmentation the better alternative. In CICS you can reduce the overall impact of segmentation I/Os by causing the segment storage areas to be placed in Main Temporary Storage (use TSMS option when executing XSPE). In XA or ESA systems this can result in a major (wall-clock) time savings since the I/O is transformed from disk I/O to virtual memory transfers (non-XA systems or those with main memory shortages should probably not use this option).

CALLed applications are not segmented and may not issue a CONVERSE when called by a segmented application.

Benefits/Costs of Segmentation

In CICS, locks and scan positions are lost every time a CONVERSE occurs. The application must reinstate any desired locks or the scan position upon return from the CONVERSE (store control information in working storage). Even though it is possible under DPPX, it not a good idea to hold a record lock between segments. This may greatly reduce concurrent use of the system.

In addition to freeing resources in CICS, each time a segment terminates, a SYNCPOINT (COMMIT) is implicitly executed. This causes a COMMIT to occur in DB2, freeing the CICS-DB2 thread and closing any open cursors. This will not change with the advent of DB2 Version 2.3's much publicized ability to keep cursors open across commits.

Combining DXFRs from a menu-driven system with segmentation allows the developer to minimize a single application's impact on the system.

Controlling Segmentation

Since segmentation is desirable in some instances and not in others, the EZESEGM word is available to allow the dynamic control of segmentation. CSP looks at the value in EZESEGM just before each CONVERSE:

If EZESEGM = 1, CSP makes the converse segmented
If EZESEGM = 0, CSP makes the converse conversational

EZESEGM is ignored in called applications. (They shouldn't issue CONVERSE anyway!)

Normally, CSP executes the same application that issued the CONVERSE upon returning to process the next segment. By DXFRing and setting a value in EZESEGTR, an application may cause a *different* transaction code to be executed upon return from the CONVERSE (even though the same application is being executed upon the return to CSP). This has useful impact only when DB2 is involved. EZESEGTR is set to a four character TRANSACTION ID pointing to the program DCBRINIT and also used to connect a plan name in the RCT. In this way, CSP applications may switch DB2 Application Plans between CONVERSEs.

Segmentation does not occur under the TEST facility. Move 1 to EZECNVCM prior to CONVERSEs to simulate the built-in SYNCPOINT/COMMIT of segmentation while testing.

Using EZESEGTR

When executing SEGMENTED transactions, the value in EZESEGTR controls which application plan is loaded for the next SQL access after a CONVERSE. To transfer from one DB2 using transaction to another DB2 using CSP application, DXFR to the other CSP application and move a different application transaction ID into EZESEGTR prior to a CONVERSE in the new application. This causes the new application plan to be loaded upon return after the user responds. The new application (the one transferred to) should not do any DB2 I-O until after it issues CONVERSE for the first time. This is because the original CICS transaction (and its application plan) is still active until the CONVERSE ends the first task. Segmented transaction IDs used must be represented in the CICS Program Control Table (PCT) and in the Resource Control Table (RCT). The transaction IDs should point to the module 'DCBRINIT' and have corresponding Invocation Parameter Group entries. This method is frequently used to switch application plans when processing DB2 and is superior to using the XFER statement.

USE OF TEMPORARY STORAGE

While waiting for the user to respond, CONVERSE stores all of the application's records in CICS Temporary Storage. System default is to use Auxiliary Temporary Storage, a cheap but potentially slow method. By using the TSMS option, an application developer may cause records held over a CONVERSE to be stored using CICS Main Temporary Storage. This is very fast but vulnerable to excess paging in memory-tight CICS regions.

24-BIT VS. 31-BIT

XA and ESA provide the capability to address data using 31 bits in an address rather than the previous limit of 24 bits. The maximum address size increases from 16 Megabytes to 2 Gigabytes with 31 bits. Older languages are "stuck" in 24-bit mode. Programs written using 24-bit capabilities cannot use 31-bit addresses. Programs using 31-bit technology can use 24-bit addresses. (To put it another way, 24 marbles fit easily into a 31-marble bag, but 31 marbles will not fit into a 24-marble bag.)

The maximum address possible with 24 bits is 16 Megabytes. Addresses that require only 24-bit addresses are said to be "below the 16 Meg line." Addresses that require 31-bit addresses are said to be "above the 16 Meg line." Under CICS, more and more of CICS's internal code is being moved "above the line." If virtual storage is plentiful, you should try to process above the line whenever possible. CSP's normal mode is to run above the line. To CALL programs written using 24-bit capabilities (such as OS/VS COBOL), use the NOTXA feature. When compiling VS COBOL II programs, the DATA(24) compile option is used when it is necessary to force data below the line. During linkage editor processing, two parameters are used in conjunction with this problem/opportunity:

AMODE Addressing mode; will the module be CALLed from a 24-bit or 31-bit program (or both)? (24,31,ANY)

RMODE Residency mode; should the module be capable of residing above the 16M line? (24,ANY)

Make sure that all your addressing requirements are synchronized. If you will be calling a 24-bit non-CSP program, specify NOTXA. If you might be called from a 24-bit program, make sure the AMODE is correct. As a general rule (but not a universal truth!), it is best to run CICS applications above the 16M line when possible.

SQL UPDATES AND SEGMENTATION

Segmented transactions raise the ugly possibility that the following sequence of events might occur:

1. User-1 looks at row (SCAN)
2. User-2 looks at same row (SCAN)
3. User-2 modifies row (UPDATE and REPLACE)
4. User-1 modifies row (UPDATE and REPLACE) based upon his or her ORIGINAL picture!

This problem is usually solved simply in one of three ways. All involve saving information at the time of the initial read, then comparing it to the record read for update before actually performing the update.

1. Storing version number data
2. Storing date, time, and ID of last update
3. Storing the entire record

Note how this method would allow User-1's application to detect that the row had changed after issuing the UPDATE (presuming that he or she looked for the condition). User-1's application could then inform the user that things have changed or perform some other wonderfully user-friendly act.

CURSORS

SQL's ability to handle SETS of data throws a monkey wrench into the processing of mere mortal traditional data processing methods. Most languages (including those at the innards of CSP) are not capable of dealing with more than one record at time. The makers of SQL (clever souls) created a logical construct called a CURSOR to take up the slack. CURSORs allow a lower-level tool to look at the result rows from a SELECT statement one row at a time as if a temporary file had been created (sometimes that's exactly what happens). The CURSOR represents the SQL statement to be executed (not unlike a VIEW), and the SQL statement's execution is controlled by the SELECT criteria within it. Some SELECT statements can be satisfied one result row at a time, without further database processing. CURSORs based upon SELECTs satisfied one row at a time are the easiest on the system since only those rows currently being looked at take up space and locks. Other SELECT statements require a degree of preprocessing before the first result row may be presented. Good illustrations are SELECT statements using DISTINCT or AVG (normally). SELECT statements that require that the result table be completed before presenting a single result row are accomplished by building a temporary table in virtual memory. The CURSOR then processes the temporary table one row at a time.

LOGICAL UNITS OF WORK AND LOCKING

Logical Units of Work (LUW) and Locking Mechanisms are built into CICS to provide some safety and integrity. Logical Units of Work are delineated by SYNCPOINTs. CICS SYNCPOINTs force DB2 COMMITs and CICS SYNCPOINT ROLLBACKs force DB2 ROLLBACKs. CSP applications may force a SYNCPOINT by issuing a CALL to EZECOMIT. CSP applications may force a SYNCPOINT ROLLBACK by CALLing EZEROLLB. The four times when SYNCPOINTs or SYNCPOINT ROLLBACKs occur are:

1. CICS SYNCPOINT (Commit LUW)
2. CICS SYNCPOINT ROLLBACK (Rollback LUW)
3. Beginning of CICS task
4. End of CICS task

Please notice that NO SYNCPOINT occurs during a DXFR. The application must force a SYNCPOINT if one is required by the logic (maybe for using Dynamic Plan Selection).

All sequential processing is ended by a SYNCPOINT. DB2 and VSAM lock data when records are read for UPDATE purposes. Locks also take place in DB2 when processing sequentially (using a cursor). The duration of locking relating to the cursor is dependent upon the ISOLATION LEVEL parameter of the BIND. Locks are not freed until a SYNCPOINT occurs (and depending upon

BIND options ACQUIRE and RELEASE maybe not then either). All locks are implicitly freed by a CONVERSE when running a SEGMENTED transaction since the task is ending.

APPLICATION PLANS AND THE RESOURCE CONTROL TABLE (RCT)

CICS provides a Resource Control Table (RCT) to define the relationship between CICS applications and DB2 Application Plans. This table is usually maintained by the CICS system programmer(s) who are (hopefully) coordinating their efforts with the DBA. DB2 Application Plans represent the system's attempt to optimize an application's SQL statements (created by BIND processing). For processing purposes, it is often advantageous to create a "static" module for a DB2-oriented application. The RCT entries connect transaction IDs that point to CSP applications (or other CICS programs) and name the Application Plan that coincides with the application. As part of the generation process, CSP stores identical timestamp information in both the Application Plan and the CSP application. If they do not match at execution time, processing halts. CICS has a limitation of one CICS task per Application Plan. This has caused several techniques to be tried in the past (discussed previously):

1. "The Application Plan That Ate Chicago"
2. XFERs and STARTs
3. EZESEGTR—CONVERSE (RETURN TRANSID)

With the installation of CICS Version 2 and DB2 Version 2 came something called "Dynamic Plan Selection." Dynamic Plan Selection causes the FIRST SQL statement encountered *after* a CICS SYNCPOINT (usually a CONVERSE when writing segmented applications) to cause the Application Plan to be loaded that matches the current SQL statement. Most CSP shops opt for using EZESEGTR and letting segmentation worry about the plan switching. (It's less work!)

DB2 Version 2 Release 3 allows creation of Package Collections. Package Collections will greatly simplify things and allow a return to "the application plan that ate Chicago" with many fewer negative effects. Big application plans have a couple of basic drawbacks: When a BIND is happening, no one can use any of the programs involved with the plan, and every time *any* module involved is changed, the BIND must be repeated, including DBRMs for all modules. Packages allow a finer level of control. Each time a program changes, its DBRM is run through a BIND PACKAGE operation and the output from the BIND is a Package in a Collection. A BIND PLAN is then run (only occasionally) using Package Collections (and DBRMs if necessary for some reason) as input to create what looks like a large application plan. When individual applications change, you perform BIND PACKAGE to modify the DBRM's Package in any Collections necessary. Any Application Plans using a Collection will automatically pick up the Package in its new form the next time the application uses that module. This reduces the length of time that BINDs require and also reduces contention problems since Packages may be bound without locking up the application plan.

STATIC VS. DYNAMIC SQL CALLS

SQL use can be separated into two groups of calls:

1. Dynamic SQL calls
2. Static SQL calls

Before an SQL statement is executed it must be parsed, checked for errors, and optimized; this process is called BINDing. Applications relying upon dynamic processing BIND each statement every time they execute (this can get kind of expensive if done a lot). Static calls assume that a BIND has been run one time and the result stored as an Application Plan (or a Collection in DB2 Version 2.3). Static SQL allows execution multiple times without incurring additional BIND processing.

This is similar to the difference between interpretive programming languages and compiled programming languages. Most of the time, static calls to DB2 greatly outperform dynamic calls. It is possible to create a "static" load module for an application allowing better use of the database as part of the generation process.

CREATING STATIC LOAD MODULES

Creating CICS/DB2 static load modules involves a few basic steps, either entirely in batch mode or in a combination of online and batch processing (until CSP Version 3.3 it is necessary to generate online at least once and save the parameters to do this).

1. Generate the application requesting a static module.
2. Put the ALF member to a sequential file.
3. Assemble the ALF member output once to expand record definitions and SQL statements properly in batch.
4. Use DB2 precompile to prepare source code for final assembly and to create DBRM (data base request module).
5. Assemble precompiled source code.
6. Link-edit the assembled module into a CICS load library paying attention to 24-bit/31-bit considerations (don't forget a PPT entry!).
7. Bind the DBRM creating an application plan.
8. Use ALFUTIL to perform new copy operation (if application is RESIDENT).
9. Use CEMT to perform a CICS NEW COPY of the static load module.

Your installation probably has procedures in place to perform this function. Figure 18.11 shows some model JCL that may be used for the ALF PUT, pre-assembly, assembly, link, and bind.

Remember, this is sample JCL, and library names and other things will have to change to make it workable for your situation.

CALLING CSP APPLICATIONS FROM CICS COBOL PROGRAMS

Command Level programs may initiate CSP programs using LINK, START, XCTL, or even RETURN TRANSID. The CICS program must pass two addresses:

1. The first address points to data area containing two eight byte sections (alfname.applname).
2. The second will map into the CALLed module's WORKING STORAGE.

```
//    JOB
//* CURRENT STATIC MODULE BEING COMPILED: PG06A$$X
//*************************************************************
//JOBLIB DD DSN=TEST.DSN220.DSNLOAD,DISP=SHR
//PROCLIB  DD  DSN=CSP.TEST.PROCLIB,DISP=SHR
//*************************************************************
//* "PUT" DB2 STATIC MODULE
//*************************************************************
//PUT     EXEC PGM=DCGBINIT
//STEPLIB  DD   DSN=CSP322.AELOAD,DISP=SHR
//DCADZGD  DD   DSN=CSP322.DZGMSG,DISP=SHR
//ALFA     DD   DSN=user.CSP.DB2WORK,DISP=SHR
//UTILALF  DD   DSN=CSP322.UTILALF.ALF,DISP=SHR
//TRGTALF  DD   DSN=user.CSP.ALF,DISP=SHR
//EZEPRINT DD   SYSOUT=*,DCB=(RECFM=VBA,LRECL=133,BLKSIZE=137)
//ALFL     DD   SYSOUT=*,DCB=(RECFM=FBA,LRECL=121,BLKSIZE=1210)
//DCAPARM DD   *
A=UTILALF.ALFBAT U=user
/*
//ALFC    DD   *
PUT MEMBER(PG06A$$X);
/*
//*************************************************************
//* PREPARE CSP MODULE
//*************************************************************
//PREPARE EXEC PREPARE,MEM=PG06A$$X,USER='SALES.TEST',DBPRE=DB2T,
//      MAC1='CSP.PROD.CSPDB2',REL2='',
//      COND.BIND=(0,LE)
//ASMH.SYSIN  DD   DSN=user.CSP.DB2WORK,DISP=SHR
//PC.DBRMLIB  DD   DSN=CSP.TEST.DBRMLIB(&MEM),DISP=SHR
//PC.SYSLIB   DD   DSN=CSP.PROD.CSPDB2,DISP=SHR
//PC.STEPLIB  DD   DSN=TEST.DSN220.DSNLOAD,DISP=SHR
//LKED.SYSLIB DD   DSN=TEST.DSN220.DSNLOAD,DISP=SHR
//LKED.SYSLMOD DD   DSN=CSP.TEST.CICSLIB(PG06A$$X),DISP=SHR
//LKED.SYSIN  DD   *
 INCLUDE SYSLIB(DSNCLI)
/*
//*
```

Figure 18.11. Compile and bind of CICS/DB2 static load module.

```
///BIND     EXEC PGM=IKJEFT01,DYNAMNBR=20
//DBRMLIB  DD DISP=SHR,DSN=CSP.TEST.DBRMLIB
//SYSTSPRT DD SYSOUT=*
//SYSPRINT DD SYSOUT=*
//SYSUDUMP DD SYSOUT=*
//*-------------------------------------------------------------
//* BIND FOR CSP APPLICATION
//*-------------------------------------------------------------
//SYSTSIN DD *
  DSN SYSTEM(DB2X)
  BIND PLAN(PG06AP) -
    MEMBER (PG06A$$X) -
    LIB('CSP.TEST.DBRMLIB') -
    ACTION (REPLACE) RETAIN -
    VALIDATE (BIND) -
    ISOLATION  (CS) -
    ACQUIRE   (USE) -
    FLAG      (I) -
    EXPLAIN   (NO) -
    RELEASE (COMMIT)
  END
//DSNENQ   DD DSN=TEST.DSN220.ENQFILE,DISP=SHR
  //
```

Figure 18.11.   Compile and bind of CICS/DB2 static load module
(continued).

It is sometimes desirable to CALL a CSP application from a CICS COBOL program. This is most often done in a menu/control program or to take advantage of CSP screen processing or edits. CSP may be called from a non-CSP program under CICS. The syntax for a CICS LINK is illustrated in Figure 18.12.

```
EXEC CICS LINK    PROGRAM      ('DCBINIT')
                  COMMAREA     (LINK-STUFF)
                  LENGTH       (LINK-STUFF-LEN)
```

Figure 18.12.   Command-level CICS LINK to CSP.

Linking to DCBINIT invokes CSP/AE. The COMMAREA (LINK-STUFF) contains two pointers (addresses). The first address points to a variable containing the ALF and application name separated by a period (ALF name and period are optional). For example ("alf1name.applname"). The second pointer points to the data area passed into the called CSP application's WORKING STORAGE. The called CSP application must be defined as a CALLED TRANSACTION, and the WORKING STORAGE record containing the passed data must be the main WORKING STORAGE referenced on the APPLICATION DEFINITION APPLICATION SPECIFICATIONS panel. The length of the passed data (LINK-STUFF-LEN) will be +8 (two fullword addresses).

CALLING NON-CSP CICS PROGRAMS FROM CSP

An application may issue a CALL to a non-CSP program as discussed in Chapter 13. The CSP call becomes a CICS LINK to the non-CSP program. The syntax for the CALL is shown in Figure 18.13. CSP passes a list of addresses into the COMMAREA (DFHCOMMAREA in COBOL) of the CALLed program pointing to each passed parameter. These addresses may then be manipulated using standard CICS and COBOL II code. See Figure 18.14.

CALL prognam wsrec (NONCSP,NOMAPS

Figure 18.13. CSP CALL to command-level CSP.

The address to the address list is passed in the COBOL program's DFHCOMMAREA. The COBOL program must then manipulate the address or list of addresses to see the data. One bit of oddness you should be aware of is that a four-byte data area is added as a FILLER field in FRONT of each field in an SQL Row Record definition passed by CSP.

```
IDENTIFICATION DIVISION.
PROGRAM- ID. PROGNAM.
ENVIRONMENT DIVISION.
DATA DIVISION.
WORKING-STORAGE SECTION.
"
LINKAGE SECTION.
01   DFHCOMMAREA.
     05   PASSED-ADDR            USAGE IS POINTER.
01   PASSED-STUFF.
     05   PASSED-FIELD-1         PIC  X(10).
     05   PASSED-FIELD-2         PIC  X(10).
     05   PASSED-FIELD-3         PIC  X(10).
     05   PASSED-FIELD-4         PIC  X(10).
"
PROCEDURE DIVISION.

***** ESTABLISH ADDRESSABILITY *****
SET ADDRESS OF PASSED-STUFF TO PASSED-ADDR.

***** PROGRAM MAY NOW USE PASSED DATA *****

"
```

Figure 18.14. Skeleton CICS/VS COBOL II program called by CSP.

CICS PERFORMANCE CONSIDERATIONS

The following is a list of things you can do to optimize processing of CSP applications under CICS (in no particular order):

1. Use SEGMENTED code.
2. Consider using multiple fixed maps with a combination of DISPLAY and CONVERSE instead of floating maps to minimize data transmissions.

3. Limit the use of MDT "on."
4. Consider "do it yourself" edits rather than CSP built-in editing if you have many users who make lots of mistakes.
5. Use DXFR rather than XFER.
6. Use CALL CREATX to initiate "response time fakery."
7. Consider making the following resident:
   segmented transactions
   large/shared map groups
   shared tables
   frequently called modules

CHAPTER 18 EXERCISES

Questions

1. What command is used to execute CSP/AD under CICS?
2. What command is used to execute CSP/AE under CICS?
3. Which CICS table must be modified to include CSP's ALFs and message files?
4. What are Invocation Parameter Groups (Parameter Entry Groups)?
5. Why are Parameter Entry Groups used?
6. What is the difference between dynamic and static SQL?
7. Why is static SQL normally preferred for production applications?
8. What CICS table entry is required for applications using static load modules that is not required for non-SQL (or dynamic SQL) applications?
9. CICS pseudoconversational processing is supported by specifying what CSP generation option?
10. What special CSP data field may be modified before a CONVERSE to get DB2 to load a different application plan upon return from the CONVERSE?

Computer Exercise
"Static" DB2-CICS-CSP Application

1. Obtain access to the CSP static load module procedure for your installation (ask your CSP administrator), or code your own using the sample in this chapter as a model.

2. Generate the application created in Chapter 12 for a CICS target system. Ask CSP to generate a static load module.
3. Use the CSP static load module procedure to process the static module, creating an executable module in the CICS load library and an application plan in DB2 for the application.
4. Transfer the application to a CICS ALF (unless you generated into a CICS ALF directly).
5. Ask the CICS system programmer to create table entries for your application:

   PCT: Entry for your primary transaction ID pointing to DCBINIT and entry for your segmented transaction ID pointing to DCBRINIT.

   PPT: Entry for your static load module.

   FCT: Entry for your ALF and CSP message file (if not already there).

   RCT: Entries connecting the transaction IDs to the application plan.

6. Do a "CEMT S PRO(applnamX) NEW" command to make sure CICS is pointing to the latest version of the static load module.
7. Create a Parameter Group Entry for your primary transaction ID naming the ALF and application.
8. Test your transaction.

19

Recent or Coming Events

Much has happened to CSP recently, and much more will be happening soon. This book has centered on the use of CSP from the TSO and CICS platforms because that is where the bulk of CSP work is being done today. CSP also works well in the VM/CMS, IMS/TM, AS/400, and DPPX environments. CSP is already a good product. With the commitment IBM has made to it, and some of the changes already in the works, it is well on its way to becoming a *great* product.

CSP PROGRAMMABLE WORKSTATION

Perhaps the most exciting thing to happen to CSP recently is the advent of the Programmable Workstation (PWS). While you have been able to generate for the personal computer environment (EZ-PREP and EZ-RUN) for some time, the PWS product gives the ability to develop on the personal computer for the first time. Those of you lucky enough to run a desktop machine with OS/2 EE may run the CSP Programmable Workstation (PWS).

The PWS allows the following:

1. Development of CSP applications on the PC
2. Development of maps using a mouse with "cut & paste" capabilities

3. Download of current applications to the desktop
4. Upload of developed/modified applications to another system (using External Source Format)

PWS does not allow the following:

1. Testing on the workstation
2. Execution on the workstation (EZ-PREP & EZ-RUN are available)

The External Source Format is used to transfer objects from the workstation to other environments. CSP/2 AD Version 1, due in August 1992, will include an integrated test facility, one-step export/upload download/import, and the ability to run generated programs on the workstation. The first cut of the PWS is useful, but without the ability to test, it can sometimes be frustrating. But, if your installation suffers from poor TSO response time, developing on the PWS may seem like a dream come true.

The current Programmable Workstation shows the direction of the future, developing on programmable workstations and executing elsewhere. The new workstation product (due August 1992) will be much better.

IMS

CSP now works in the IMS/TM (IMS-DC) environment through a new product called CSP/370 Runtime Services. CSP/AD and CSP/AE do not run under IMS. CSP/AD in other environments is now able to generate applications for IMS by creating VS CO-BOL II code. First, this is not the prettiest code around, but it gets the job done. Second, you are *not* to modify this code. If you feel the need to enhance an application's functionality, issue a CALL to another "normal" program to do that work.

The VS COBOL II code is run through normal compile and link procedures to create executable (load) modules to run under IMS. The generation process will even build MFS source and the compilation JCL for you based upon models supplied with CSP and modifiable at your installation. You will need to supply your own MFS assembly, LINK, and BIND JCL.

To execute, the CSP/370 Runtime Services must be available in the concatenation of load libraries available to IMS. It is important to remember that the execution environment is pure COBOL. Main transactions may be run under IMS as MPPs. Batch applications may be executed as MPPs, BMP applications, or MVS batch applications. CSP/370 Runtime Services needs to be available for batch jobs too.

VS COBOL II FOR BATCH APPLICATIONS

CSP/AD will also create VS COBOL II source for batch applications under MVS. The process is simple: generate COBOL code for batch and then compile it normally (making sure the CSP/370 Runtime Services modules are available). Then execute as you would a normal batch program (again, making sure that CSP/370 Runtime Services modules are available).

SCREEN GENERATION

The screen definition portion of CSP may be mostly replaced with IBM's SDF II product, the official screen painter of SAA. This is a good facility that allows you to "paint" screens once under ISPF (either under TSO or VM), then generate the maps for CSP, CICS/BMS, IMS/MFS, Dialog Manager, or GDDM.

Screens may also be developed under OS/2 using EASEL when working on the PWS.

FUTURE DIRECTIONS

CSP is part of IBM's overall System Application Architecture (SAA) blueprint (no pun intended). IBM will continue to improve and extend CSP to complement the CASE tools being installed as part of AD/Cycle completely.

CSP/370AD Version 4 for MVS/ESA is scheduled for August 28, 1992, along with CSP/370 Runtime Services Version 2, CSP/2 AD for OS/2 Version 1, and CSP/2 Runtime Services for OS/2 Version 1. It is intended that CSP remain an Application Generator; it is not meant to replace COBOL or PL/I, but CSP/AD Version 4 will generate VS COBOL II code for CICS/MVS, CICS/

ESA, CICS/OS2, IMS/BMP, IMS/TM, MVS BATCH, and TSO. This COBOL code can be compiled and linked into ordinary load module libraries. CSP Runtime Services will also be required at execution time. CSP/AE will not be necessary to run programs with the new release.

A partial list of goodies in the new release also includes:

Expanded arithmetic and conditional statements
    Mixed ANDs and ORs for IF and WHILE
    IF and TEST use same options
    Parentheses in calculations
Map support for DB2 dates/times
SET capability for color and extended attributes
Generalized date and time routines
Century in date masks, and much more
CICS user ID availability
CICS temporary storage access
CICS printing
Direct import/export between MSLs
More focused Where-Used ability
New CSP/AD batch utility capabilities
    Associates list
    Copy with replace
    Move
    Copy and Delete with associates
    Generic keys
Ability to CALL the ALFUTIL from a program
Ability to make application with associates Resident
More EZE words
Environment check within application using a new EZE word

As you can see, between the new capabilities and COBOL generation the next release is something to look forward to. CSP will continue to grow in power and flexibility every year. Look for the most exciting developments in the PWS (Programmable Workstation) environment.

CSP JCL and Command Summary

MVS JCL FOR MSL CREATION

```
//xxx       JOB ————
//mslstep   EXEC PGM=IDCAMS
//SYSPRINT  DD SYSOUT=*
//SYSIN     DD *
 DEFINE CLUSTER (NAME(your.msl.name)              -
                 VOL(xxxxxx)                       -
                 CYL(pri sec)                      -
                 KEYS(57 0)                         -
                 RECSZ(avg max)                    -
                 INDEXED                            -
                 )                                  -
        DATA    (NAME(your.msl.name.data))         -
        INDEX   (NAME(your.msl.name.index))
/*
//
```

It is best if your MSLs share a common record size (several shops use RECSZ(1024 4092))

Minimum record size (avg) allowed by CSP is 292

Maximum record size (max) allowed by CSP is 32,760

Be sure the MSL is allocated prior to attempting to reference it in CSP

MVS JCL FOR ALF CREATION

```
//xxx      JOB ─────────
//alfstep  EXEC PGM=IDCAMS
//SYSPRINT DD SYSOUT=*
//SYSIN    DD *
 DEFINE CLUSTER (NAME(your.alf.name)                    -
                 VOL(xxxxxx)                            -
                 CYL(pri sec)                           -
                 KEYS(57 0)                             -
                 RECSZ(avg max)                         -
                 INDEXED                                -
                 )                                      -
        DATA    (NAME(your.alf.name.data))             -
        INDEX   (NAME(your.alf.name.index))
/*
//
```

Minimum record size allowed (avg) by CSP is 272
Maximum record size allowed (max) by CSP is 32,528

CSP/AD COMMANDS USED ON ALL DISPLAYS

COMMAND	FUNCTION
?	Retrieve previous command(s) (up to 7)
/	Same as FIND
+nnn	Scroll towards end of data nnn lines
-nnn	Scroll towards beginning of data nnn lines
ALARM n	Causes alarm to sound if n seconds elapses between user inputs
BOTTOM	Scroll to bottom of current list/data
CANCEL	End current function, ignore changes

CHANGE | CHG | C   oldtxt newtxt
                            [All | First | Last | Next | Prev | Rest]
                            [ASis]
                            Alter text strings

EXIT   Leave current function and save changes

FIND | LOCATE | /   textstrg
                            [First | Last | Next | Prev] [ASis]
                            Find or Locate text string

LOCATE   Same as FIND

MAPID   Toggles display of CSP/AD map id in upper-left
                            corner of each screen

MESSAGE   Displays the requested CSP/AD message (see
                            message file utility later in this course)

QUIT   Return to CSP/AD Facility Selection or List
                            Processor display, leave current function and
                            ignore changes

RETURN   Save and return to CSP/AD Facility Selection or
                            List Processor display

SCROLL   [Left(Page|Half|Max|nn)]   [Right(Page|Half|Max|nn)]
              [Up(Page|Half|Max|nn)]   [Down(Page|Half|Max|nn)]
                         Set scroll values

TOP   Scroll to first line

CSP/AD COMMANDS ONLY FOR MAP DISPLAYS

COMMAND   **FUNCTION**

ATTRIBUTE   Changes attributes for current field

CODES   [C(x)]   [V(x)]   [(S)]
                            Change/display Constant, Variable, and Spacer
                            codes

COPY [After|BEfore]   Copy line / block defined by FROM-TO

DELETE   [n]   Deletes specified number of lines (includes
                            current)

FROM	Marks the beginning of a block
INPUT [[n] [After|Before]	Insert lines into map
LSHIFT [n]	Shift line / block left n columns
MOVE [After|BEfore]	Move line / block
REPEAT [n]	Repeat lines defined by FROM-TO
RSHIFT [n]	Shift line / block right n columns
SAVE	Save changes, do not leave display
TEST [FILL(x)]	Display map with fill chars. in PROT/ASKIP fields
TO	Mark the end of a block

OTHER CSP/AD "COMMAND LINE" COMMANDS

COMMAND	FUNCTION
COPYLIST	Copy all members on the current list from read-only MSLs to the read/write MSL
MOVE	Change storage during testing
RESET	Ignore and remove "line" commands, remove COLUMN and MASK settings
SHOW	Display storage during test

CSP/AD "LINE COMMAND" COMMANDS

LINE CMD	FUNCTION
/	Locate, shifts current line to top of display
COL	Show ruler line (not part of data)
MSK	Show/update mask line
A	Copy/Move lines After this line
B	Copy/Move lines Before this line

C	Marks line to copy
Cn	Marks n lines to be copied (beginning with this one)
CC	Marks block of lines to be copied
D	Delete this line
Dn	Delete n lines (beginning with this one)
DD	Mark block of lines for deletion
I	Insert a new line
In	Insert n new lines
M	Marks line to move
Mn	Marks n lines to be moved (beginning with this one)
MM	Marks a block of lines to be moved
R	Repeat this line
Rn	Repeat this line n times
RR	Repeat block of lines once
RRn	Repeat block of lines n times

CSP/AD LINE COMMANDS USED ONLY IN PROLOGUE AND PROCESS DEFINITION

LINE CMD	FUNCTION
>	Shift current line right 1 position
>n	Shift current line right n positions
>>	Shift current block of lines right 1 position
>>n	Shift current block of lines right n positions
<	Shift current line left 1 position
<n	Shift current line left n positions
<<	Shift current block of lines left 1 position
<<n	Shift current block of lines left n positions

CSP PROCESSING STATEMENTS

%GET [appl.]procname

object.result = operand1 operator operand2 [. . .] [(r];

CALL subrtn [arg1,arg2,...,argn] [([NOMAPS] [NONCSP] [REPLY]];

CALL CREATX recdname,tdqq[,termid];

DXFR cspappl|progname|EZEAPP [ws_rec_nam] [NONCSP];

FIND itemval tablnam[.colnam] true_stmt_group,false_stmt_group;

IF condition;
[AND|OR condition2 ...];
    resultoption;

ELSE;
    resultoption;

END;

IF recordname IS|NOT DED|DUP|EOF|ERR|FUL|HRD|LOK|NRF|UNQ;

IF mapname.mapfld IS|NOT   [BLANKS|BLANK] [NULLS|NULL]
        [DATA] [CURSOR] [MODIFIED];

IF mapname IS|NOT MODIFIED;

IF EZEAID IS | NOT   [PF1|PF2|...|PF24|PF] [PA1|PA2|PA3|PA]
        [ENTER] [BYPASS];

IF sqlrec.fld IS|NOT   NULL|TRUNC;

MOVE operand1 TO operand2;

MOVEA operand1 TO array2 FOR operand3;

RETR itemval tablnam[.colnam] returned_item [return_tab_col];

SET recordname EMPTY;

SET SQL sqlrec.item NULL;

SET recordname SCAN;

SET mapname.item [NORMAL|DEFINED] [CURSOR] [FULL];

SET mapname [PAGE] [ALARM] [CLEAR|EMPTY];

SET mapname.item [CURSOR] [FULL] [MODIFIED]
        [BRIGHT|DARK] [PROTECT|AUTOSKIP];

TEST operand condition [trueoption],[falseoption];

WHILE condition;
     resultoption;
[AND|OR condition2 ...];
END;

XFER transid|cspappl|progname|EZEAPP [ws_rec_nam] [NONCSP]

CSP PROCESS OPTIONS

PROCESS OPTION	DESCRIPTION
ADD	Insert new record/row
CLOSE	End sequential process
CONVERSE	Send map to screen and receive response
DELETE	Delete record/row
EXECUTE	Code not performing I/O
INQUIRY	Read record/row
REPLACE	Rewrite record/row
SCAN	Read next record/row
SCANBACK	Read previous VSAM KSDS record/row
SETINQ	Begin SQL sequential read-only process
SETUPD	Begin SQL sequential read-update process
SQLEXEC	Developer-coded SQL statement
UPDATE	Read record/row, lock for update

Sample CSP Application Documentation

MSL CONCATENATION SEQUENCE

MSL ACCESS   MSL FILE
ORDER   NAME

READ/WRITE MSL: 1   MYRWMSL

READ-ONLY MSL(S):   2   ROMSL1
                          3
                          4
                          5
                          6

APPLICATION NAME: TZ01APP
TYPE OF APPLICATION: MAIN TRANSACTION
WORKING STORAGE: TZ01W
MAP GROUP NAME: TZ01G
HELP MAP GROUP NAME:
HELP MAP PF KEY:
BYPASS EDIT PF KEYS: 3
PF1-12=PF13-24:   YES
ALLOW IMPLICITS: NO
MESSAGE FILE: XYZ-
MSL: MYRWMSL

PROLOGUE   THIS IS AN EXAMPLE APPLICATION

```
TZ01MAIN    ****************************************************************************
            *           MAINLINE ROUTINE FOR TX21A                   MYRWMSL       *
            ****************************************************************************
            OPTION -    EXECUTE                        OPTION                                          2

                        :/* *** INITIALIZATION STUFF ***                                              5
                        SET EMPMAP EMPTY;                                                              7
                        SET TZ01W EMPTY;                                                               8
                        MOVE 1 TO EZEFEC;          /* *** ALLOWS PROG TO CATCH DB2 ERRORS ***         9
                        MOVE 1 TO EZECNVCM;        /* *** SIMULATES PSEUDO-CONV FOR TSO              10
                                                   /* TESTS ***                                      11
                                                                                                     12
                        :/* *** "REAL LOGIC" ***                                                     13
                        MOVE 25 TO EZEMNO;                                                            14
                        WHILE QUITFLD NE 'QUIT';                                                      15
                          AND EZEAID NOT PF3;                                                         16
                          PERFORM TZ01SEND;                                                           17
                        END;                                                                          18
                        EZECLOS;                   /* NOTE THAT COMMENTS MAY FOLLOW SEMI-COLON        19
                                                   /* END OF TRANSACTION                             20
```

ADDITIONAL PERFORMED PROCESSES

TZ01SEND

```
           ************************************************************************
           *              MAP SEND FOR TX21A                    MYRWMSL  *
           ************************************************************************
           OPTION -   CONVERSE EMPMAP              MAP

22             IF EMPMAP NOT MODIFIED;
25                 MOVE 'HEY, YOU GOTTA ENTER SOMETHING' TO EMPMAP.EZEMSG;
               ELSE;
27                 IF EMPID IS MODIFIED;
28                     PERFORM TZ01GETREC;
29                 ELSE;
30                     IF FNAME IS MODIFIED;
31                     OR LNAME IS MODIFIED;
32                     OR EMPPAY IS MODIFIED;
33                         PERFORM TZ01READUPD;
34                         IF EZESQCOD = 0;
35                             IF EMPNO = WS-CHECK-EMPNO;
36                             AND FIRSTNME = WS-CHECK-FIRSTNME;
37                             AND LASTNAME = WS-CHECK-LASTNAME;
38                             AND SALARY = WS-CHECK-SALARY;
39                                 MOVE FNAME TO FIRSTNME;
40                                 MOVE LNAME TO LASTNAME;
41                                 MOVE EMPPAY TO SALARY;
42                                 PERFORM TZ01REWRITE;
43                                 MOVE 28 TO EZEMNO;
```

```
47          ELSE;
48            SET EMPID MODIFIED,BRIGHT;
49            MOVE EMPNO TO EMPID;
50            MOVE FIRSTNME TO FNAME;
51            MOVE LASTNAME TO LNAME;
52            MOVE SALARY TO EMPPAY;
53            MOVE 27 TO EZEMNO;
54          END;
55        ELSE;
56          MOVE 22 TO EZEMNO;
57        END;
58      END;
59    END;
60
```

```
                  ************************************************************
                                                                MYRWMSL        *
                  ************************************************************
62  TZ01GETREC

65    MOVE EMPID TO EMPNO;
66    SET EMPMAP EMPTY;
67    MOVE EMPNO TO EMPID;
```

```
SQL STATEMENT MODIFIED:                              NO
EXECUTION TIME STATEMENT BUILD:                      NO
SINGLE ROW SELECT:                                   NO

SELECT                                                           70
    EMPNO, FIRSTNME, LASTNAME, SALARY                           71
INTO                                                            72
    :EMPNO, :FIRSTNME, :LASTNAME, :SALARY                       73
FROM                                                            74
    TSOUSR.EMP EMP                                              75
WHERE                                                           76
    EMPNO = :EMPID                                              77
/*** INSERT ORDER BY CLAUSE HERE **                             78
                                                                79
IF EZESQCOD = 0;                                                80
    MOVE EMPNO TO WS-CHECK-EMPNO;                               81
    MOVE LASTNAME TO WS-CHECK-LASTNAME;                         82
    MOVE FIRSTNME TO WS-CHECK-FIRSTNME;                         83
    MOVE SALARY TO WS-CHECK-SALARY;                             84
    MOVE FIRSTNME TO FNAME;
    MOVE LASTNAME TO LNAME;
    MOVE SALARY TO EMPPAY;
ELSE;
    IF EZESQCOD = +100;     /* * RECORD NOT FOUND *
        MOVE 21 TO EZEMNO;
    ELSE;                   /* * SERIOUS ERROR *
        MOVE 23 TO EZEMNO;
    END;

END;
```

TZ01READUPD

```
*********************************************************************
*                                                          MYRWMSL  *
*********************************************************************
OPTION -    UPDATE   TZ01REC          RECORD          EZERTN

                                                                    86

                                                                    89

            SQL STATEMENT MODIFIED:           NO
            EXECUTION TIME STATEMENT BUILD:   NO

            SELECT
              EMPNO, FIRSTNME, LASTNAME, SALARY
            INTO
              :EMPNO, :FIRSTNME, :LASTNAME, :SALARY
            FROM
              TSOUSR.EMP EMP
            WHERE
              EMPNO = :EMPID
            FOR UPDATE OF
              EMPNO, FIRSTNME, LASTNAME, SALARY
```

TZ01REWRITE   ***   91
                *   MYRWMSL *
                ****

                OPTION -   REPLACE TZ01REC   RECORD   EZERTN   94

                      SQL STATEMENT MODIFIED:
                      UPDATE OR SETUPD PROCESS NAME: NO

                      UPDATE
                         TSOUSR.EMP
                      SET
                         EMPNO=:EMPNO,
                         FIRSTNME=:FIRSTNME,
                         LASTNAME=:LASTNAME,
                         SALARY=:SALARY
                         WHERE CURRENT OF EZE_CURSOR_NNN

NO GROUPS FOR THIS APPLICATION

APPLICATION STRUCTURE

NAME	LVL	OPTION	OBJECT	ERROR	DESCRIPTION	LINE
TZ01MAIN	001	EXECUTE			MAINLINE ROUTINE FOR TX21A	2
TZ01SEND	002	CONVERSE	EMPMAP		MAP SEND FOR TX21A	22
TZ01GETREC	003	INQUIRY	TZ01REC	EZERTN		62
TZ01READUPD	003	UPDATE	TZ01REC	EZERTN		86
TZ01REWRITE	003	REPLACE	TZ01REC	EZERTN		91

APPLICATION NAME: TZ01APP   CSP/AD   DATE: 04/02/92   TIME: 15:34:36   PAGE: 00007
GENERATION OPTIONS

NO GENERATION OPTIONS SAVED

RECORD NAME: TZ01W

ORGANIZATION: WORKING STORAGE
LENGTH IN BYTES: 52
DEFAULT SCOPE: LOCAL
MSL: MYRWMSL

PROLOGUE

THIS WS RECORD WAS CREATED FOR BREAKING DOWN THE PART NUMBER
FOR SPECIAL EDIT PURPOSES.

NAME	LVL	OCCURS	TYPE	SCOPE	LENGTH	DEC	BYTES	START	MSL	DESCRIPTION
WS-BROWSE-KEYS	10		CHA	LOCAL	12		12	1		
WS-BROWSE-FWD-KEY	15		CHA	LOCAL	6		6	1		
WS-BROWSE-BWD-KEY	15		CHA	LOCAL	6		6	7		
WS-SUB	10		BIN	LOCAL	4		2	13		
WS-CHECK-FIELDS	10		CHA	LOCAL	38		38	15		
WS-CHECK-EMPNO	15		CHA	LOCAL	6		6	15		
WS-CHECK-FIRSTNME	15		CHA	LOCAL	12		12	21		
WS-CHECK-LASTNAME	15		CHA	LOCAL	15		15	33		
WS-CHECK-SALARY	15		PACK	LOCAL	9	2	5	48		

RECORD NAME: TZ01REC

ORGANIZATION: SQL ROW
LENGTH IN BYTES: 54
DEFAULT SCOPE: LOCAL
MSL: MYRWMSL

SQL TABLE NAME(S)

TABLE NAME   TABLE LABEL

TSOUSR.EMP   EMP

DEFAULT SELECTION CONDITIONS
SELECT
   EMPNO, FIRSTNME, LASTNAME, SALARY
INTO
   :EMPNO, :FIRSTNME, :LASTNAME, :SALARY
FROM
   TSOUSR.EMP EMP
WHERE
/*** INSERT DEFAULT SELECT CONDITIONS HERE **
   EMPNO = :EMPID

NO PROLOGUE

NAME / DESCRIPTION	TYPE	SCOPE	LENGTH	DEC	BYTES	START	R/O	KEY	SQL DATA CODE/ SQL COLUMN NAME	MSL
EMPNO EMPNO	CHA	LOCAL	6		6	5	NO	NO	453 EMPNO	
FIRSTNME FIRSTNME	CHA	LOCAL	12		12	15	NO	NO	449 FIRSTNME	
LASTNAME LASTNAME	CHA	LOCAL	15		15	31	NO	NO	449 LASTNAME	
SALARY SALARY	PACK	LOCAL	9	2	5	50	NO	NO	485 SALARY	

400

APPLICATION NAME: TZ01APP   CSP/AD   DATE: 04/02/92 TIME: 15:34:36 PAGE: 00010
          MAPGROUP: TZ01G   MSL:

                       MAP GROUP NAME: TZ01G   (NO MAPGROUP MEMBER DEFINED)

                       NUMBER OF SUPPORTED DEVICES 1

TOTAL MAPS   1   * = SUPPORTED,   S = SIZE ERROR

MAP: NAME   LINE COLUMN DEPTH WIDTH MSL   3278-2
     EMPMAP   1   1   24   80 MYRWMSL   *

APPLICATION NAME: TZ01APP   CSP/AD   DATE: 04/02/92   TIME: 15:34:36   PAGE: 00011
        MAPGROUP: TZ01G   MAP: EMPMAP   MSL: MYRWMSL

SPECIFICATION FOR FOLLOWING DEVICE TYPES

3278-2

MAP CHARACTERISTICS

    MAP SIZE: DEPTH => 24
                                             WIDTH => 80
    MAP POSITION: LINE => 1
                                             COLUMN=> 1

    CURSOR FIELD   => EMPID

FIELD AND ATTRIBUTE DEFINITION

        CONSTANT FIELD CODE: #
        VARIABLE FIELD CODE: ^
        SPACER CHARACTER: /

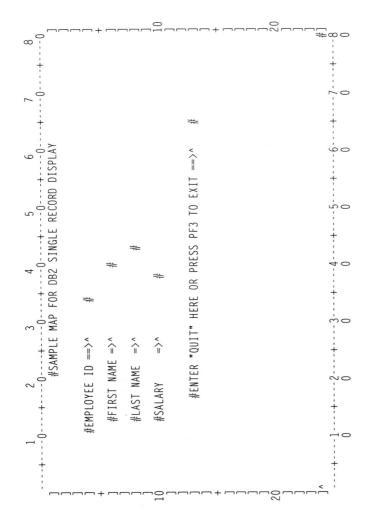

```
#SAMPLE MAP FOR DB2 SINGLE RECORD DISPLAY


  #EMPLOYEE ID ==>^          #

   #FIRST NAME =>^        #

   #LAST NAME  =>^            #

   #SALARY     =>^      #


        #ENTER "QUIT" HERE OR PRESS PF3 TO EXIT ==>^     #
```

APPLICATION NAME: TZ01APP   DATE: 04/02/92   TIME: 15:34:36   PAGE: 00012
        MAPGROUP: TZ01G   CSP/AD   MAP: EMPMAP   MSL: MYRWMSL

FIELD ATTRIBUTE CHARACTERS

LINE COLUMN   FIELD NAME   ATTRIBUTES

 4   27   EMPID ....   UNPROTECT BRIGHT
 6   27   FNAME ....   UNPROTECT
 8   27   LNAME ....   UNPROTECT
10   27   EMPPAY ....   UNPROTECT
13   60   QUITFLD ....   UNPROTECT
24   1   EZEMSG ....   ASKIP BRIGHT

MAP ITEM DEFINITIONS

	OCCURS	TYPE/ERR	LENGTH	DEC	JUS	FOLD	FIL	ZERO	MIN/ERR	SIGN	SEP	MONY	REQ/ERR	EDIT	RTN/ERR	MIN	MAX	ERR	HEX
EMPID		CHA	6						EMPLOYEE ID										
					LEF	MAP	N	NO		NO	NO	NO	NO						NO
FNAME		CHA	12						FIRST NAME										
					LEF	MAP	N	NO		NO	NO	NO	NO						NO
LNAME		CHA	15						LAST NAME										
					LEF	MAP	N	NO		NO	NO	NO	NO						NO
EMPPAY		NUM	10	2					SALARY										
					RIG	MAP	N	NO		NO	NO	NO	NO			BELOW		0006	NO
										RANGE	MIN:	00500.00				MAX:	50000.00		
QUITFLD		CHA	4																
					LEF	MAP	N	NO	04	NO	NO	NO	NO						NO
EZEMSG		CHA	78		NO	MAP	N	NO		NO	NO	NO	NO						

APPLICATION NAME: TZ01APP
APPLICATION CROSS REFERENCE

CSP/AD

DATE: 04/02/92 TIME: 15:34:36 PAGE: 00013

MEMBER NAME	MSLTYPE	USAGE	REFERENCE	REFTYPE	PAGE	LINE	MOD	SCOPE	TYPE	BYTES DEC
EMPID	ITEM	DEFINED EMPMAP	TZ01G	MAP	012	00030		LOCAL	CHA	00006
		USED	TZ01SEND	PROC	003	00048				
		USED	TZ01SEND	PROC	003	00049	*			
		USED	TZ01GETREC	PROC	003	00065				
		USED	TZ01GETREC	PROC	003	00067	*			
		USED	TZ01GETREC	PROC	003	00068				
		USED	TZ01READUPD	PROC	004	00089				
		USED	TZ01REC	RECD	009					
EMPMAP	MAP	DEFINED	TZ01G		012	00008	*			
		USED	TZ01MAIN	PROC	002	00025				
		USED	TZ01SEND	PROC	003	00027				
		USED	TZ01SEND	PROC	003	00028				
		USED	TZ01SEND	PROC	003	00066	*			
		USED	TZ01GETREC							
EMPNO	ITEM	USED	TZ01SEND	PROC	003	00038				
		USED	TZ01SEND	PROC	003	00049				
	ITEM	DEFINED	TZ01REC	RECD	009			LOCAL	CHA	00006
		USED	TZ01GETREC	PROC	003	00065	*			
		USED	TZ01GETREC	PROC	003	00067				
		USED	TZ01GETREC	PROC	003	00068	*			
		USED	TZ01READUPD	PROC	003	00071				
		USED	TZ01REC	RECD	004	00089	*			
		USED	TZ01REWRITE	PROC	005	00094				

```
ITEM       DEFINED EMPMAP    TZ01G      MAP  012            LOCAL  NUM  00010  02
EMPPAY             USED TZ01SEND        PROC 003  00035
                   USED TZ01SEND        PROC 003  00044
                   USED TZ01SEND        PROC 003  00052 *
                   USED TZ01GETREC      PROC 003  00077 *

EZEAID             USED TZ01MAIN        PROC 002  00017

EZECLOS            USED TZ01MAIN        PROC 002  00020

EZECNVCM           USED TZ01MAIN        PROC 002  00011 *

EZEFEC             USED TZ01MAIN        PROC 002  00010 *

EZEMNO             USED TZ01MAIN        PROC 002  00015 *
                   USED TZ01SEND        PROC 003  00046 *
                   USED TZ01SEND        PROC 003  00053 *
                   USED TZ01SEND        PROC 003  00056 *
                   USED TZ01GETREC      PROC 003  00080 *
                   USED TZ01GETREC      PROC 003  00082 *
```

CSP/AD   DATE: 04/02/92 TIME: 15:34:36 PAGE: 00014

MEMBER NAME	MSLTYPE	USAGE	REFERENCE	REFTYPE	PAGE	LINE	MOD	SCOPE	TYPE	BYTES	DEC
EZEMSG	ITEM	DEFINED	EMPMAP TZ01G	MAP	012			LOCAL	CHA	00078	
		USED	TZ01SEND	PROC	003	00028	*				
EZERTN		USED	TZ01GETREC	PROC	003	00068					
		USED	TZ01READUPD	PROC	004	00089					
		USED	TZ01REWRITE	PROC	005	00094					
EZESQCOD		USED	TZ01SEND	PROC	003	00037					
		USED	TZ01GETREC	PROC	003	00070					
		USED	TZ01GETREC	PROC	003	00079					
FIRSTNME	ITEM	USED	TZ01SEND	PROC	003	00039					
		USED	TZ01SEND	PROC	003	00042	*				
		USED	TZ01SEND	PROC	003	00050					
	ITEM	DEFINED	TZ01REC	RECD	009			LOCAL	CHA	00012	
		USED	TZ01GETREC	PROC	003	00068	*				
		USED	TZ01GETREC	PROC	003	00073					
		USED	TZ01GETREC	PROC	003	00075					
		USED	TZ01READUPD	PROC	004	00089	*				
		USED	TZ01REC	RECD	009	00094					
		USED	TZ01REWRITE	PROC	005	00094					
FNAME	ITEM	DEFINED	EMPMAP TZ01G	MAP	012			LOCAL	CHA	00012	
		USED	TZ01SEND	PROC	003	00033					
		USED	TZ01SEND	PROC	003	00042					
		USED	TZ01SEND	PROC	003	00050	*				
		USED	TZ01GETREC	PROC	003	00075	*				
LASTNAME	ITEM	USED	TZ01SEND	PROC	003	00040					
		USED	TZ01SEND	PROC	003	00043	*				
		USED	TZ01SEND	PROC	003	00051					

```
ITEM          DEFINED TZ01REC        RECD 009 00068 *   LOCAL CHA 00015
              USED    TZ01GETREC     PROC 003 00072
              USED    TZ01GETREC     PROC 003 00076
              USED    TZ01READUPD    PROC 004 00089 *
              USED    TZ01REC        RECD 009
              USED    TZ01REWRITE    PROC 005 00094

LNAME   ITEM  DEFINED EMPMAP  TZ01G  MAP  012 00034    LOCAL CHA 00015
              USED    TZ01SEND       PROC 003 00043
              USED    TZ01SEND       PROC 003 00051 *
              USED    TZ01SEND       PROC 003 00076 *
              USED    TZ01GETREC     PROC 003

QUITFLD ITEM  USED    TZ01MAIN       PROC 002 00016    LOCAL CHA 00004
        ITEM  DEFINED EMPMAP  TZ01G  MAP  012

SALARY  ITEM  USED    TZ01SEND       PROC 003 00041
```

MEMBER NAME	MSLTYPE	USAGE	REFERENCE	REFTYPE	PAGE	LINE	MOD	SCOPE	TYPE	BYTES	DEC
	ITEM	USED	TZ01SEND	PROC	003	00044 *					
		USED	TZ01SEND	PROC	003	00052					
		DEFINED	TZ01REC	RECD	009			LOCAL	PACK	00005	02
		USED	TZ01GETREC	PROC	003	00068 *					
		USED	TZ01GETREC	PROC	003	00074					
		USED	TZ01GETREC	PROC	003	00077					
		USED	TZ01READUPD	PROC	004	00089 *					
		USED	TZ01REC	RECD	009						
		USED	TZ01REWRITE	PROC	005	00094					
TZ01GETREC	PROC	DEFINED			003						
		USED	TZ01SEND	PROC	003	00031					
TZ01MAIN	PROC	DEFINED			002						
TZ01READUPD	PROC	DEFINED			004						
		USED	TZ01SEND	PROC	003	00036					
TZ01REC	RECD	DEFINED			009						
		USED	TZ01GETREC	PROC	003	00068					
		USED	TZ01READUPD	PROC	004	00089					
		USED	TZ01REWRITE	PROC	005	00094					
TZ01REWRITE	PROC	DEFINED			005						
		USED	TZ01SEND	PROC	003	00045					
TZ01SEND	PROC	DEFINED			003						
		USED	TZ01MAIN	PROC	002	00018					
TZ01W	RECD	DEFINED			008						
		USED	TZ01MAIN	PROC	002	00009 *					

Name		Reference	Location		Type	Size
WS-BROWSE-BWD-KEY	ITEM	DEFINED TZ01W	RECD 008	LOCAL	CHA	00006
WS-BROWSE-FWD-KEY	ITEM	DEFINED TZ01W	RECD 008	LOCAL	CHA	00006
WS-BROWSE-KEYS	ITEM	DEFINED TZ01W	RECD 008	LOCAL	CHA	00012
WS-CHECK-EMPNO	ITEM	USED TZ01SEND	PROC 003 00038			
		USED TZ01GETREC	PROC 003 00071 *			
	ITEM	DEFINED TZ01W	RECD 008	LOCAL	CHA	00006
WS-CHECK-FIELDS	ITEM	DEFINED TZ01W	RECD 008	LOCAL	CHA	00038
WS-CHECK-FIRSTNME	ITEM	USED TZ01SEND	PROC 003 00039			
		USED TZ01GETREC	PROC 003 00073 *			
	ITEM	DEFINED TZ01W	RECD 008	LOCAL	CHA	00012
WS-CHECK-LASTNAME	ITEM	USED TZ01SEND	PROC 003 00040			
		USED TZ01GETREC	PROC 003 00072 *			

MEMBER NAME	MSLTYPE	USAGE	REFERENCE	REFTYPE	PAGE	LINE	MOD	SCOPE	TYPE	BYTES	DEC
	ITEM	DEFINED	TZ01W	RECD	008			LOCAL	CHA	00015	
WS-CHECK-SALARY											
	ITEM	USED	TZ01SEND	PROC	003	00041					
		USED	TZ01GETREC	PROC	003	00074	*				
	ITEM	DEFINED	TZ01W	RECD	008			LOCAL	PACK	00005	02
WS-SUB											
	ITEM	DEFINED	TZ01W	RECD	008			LOCAL	BIN	00002	

END OF APPLICATION CROSS REFERENCE

Appendix C

"EZE" Words

"EZE" WORD	DESCRIPTION
EZEAID	One character field containing a value which represents the function key pressed by the user
EZEAPP	Eight byte character field that contains an application name, this name is used by DXFR and XFER to dynamically control the transfer-to name
EZECLOS	Routine that CLOSES a CSP/AD application (default for last process in application)
EZECNVCM	Move a "1" to this one byte field to cause
EZECOMIT	to execute automatically after every display caused by a CONVERSE
EZECOMIT	Routine that causes SQL COMMIT WORK under TSO or VM, SYNCPOINT under CICS, CHKP under DL/I, and DTMS COMMIT under DPPX
EZEDAY	Five byte field containing the Julian date "YYDDD"
EZEDTE	Six byte field containing the Gregorian date in the form "YYMMDD"
EZEFEC	Move a "1" to this field in SQL applications—If this one byte field's value is "0" (the default) the

	application is terminated by hard I-O (and SQL) errors, If the value is "1" the application continues to process and is responsible for handling the error situation
EZEFLO	Routine that causes current statement group or sub process to GO TO the FLOW portion of the main process
EZEMNO	Two byte binary field used to obtain messages from the message file
EZEMSG	Field containing error message
EZEOVERS	One byte numeric field set to "1" when arithmetic overflow occurs
EZERCODE	Four byte binary field used to set a JCL return code
EZEROLLB	Routine that causes SQL ROLLBACK under TSO or VM, SYNCPOINT ROLLBACK under CICS, ROLB under DL/I, and DTMS RESET under DPPX
EZERT8	Eight character field containing file status values after I-O
EZERTN	Routine that exits from current statement group or process, equivalent to EZEFLO if executed from main process
EZESEGM	One byte switch used to turn segmentation "on" or "off"
EZESEGTR	Eight byte field containing the CICS transaction id to be activated when returning from the next CONVERSE, this field is meaningful only when segmentation is in effect and useful only when DB2 is involved
EZESQCOD	Contains the four byte binary SQLCODE, value 0 means everything is OK, positive values are warnings/informatory, negative values represent errors (0 = ok, 100 = nrf/enddata, xx0 = error)
EZESQRD3	Contains SQLERRD(3), a four byte binary field containing the number of rows processed by SQL UPDATE or DELETE

EZESQRRM   Seventy character field containing text for the current
                    SQL error

EZETIM   Eight byte field containing current time in the form
                    "HH:MM:SS"

EZETST   Two byte binary field modified by most statements
                    that manipulate or interrogate arrays

EZEUSR   Eight byte field (ten on OS/400) contains different
                    information depending on the environment:

TSO	User ID
CICS	Terminal ID
CMS	Logon ID
DPPX	User ID
OS/400	Job User Profile Name

Index